Cambridge Tracts in Mathematics

No. 57

METRIC SPACES

METRIC SPACES

E. T. COPSON

Formerly Regius Professor of Mathematics in the University of St Andrews

CAMBRIDGE UNIVERSITY PRESS
Cambridge, New York, Melbourne, Madrid, Cape Town, Singapore, São Paulo, Delhi

Cambridge University Press
The Edinburgh Building, Cambridge CB2 8RU, UK

Published in the United States of America by Cambridge University Press, New York

www.cambridge.org
Information on this title: www.cambridge.org/9780521047227

First published 1968
Reprinted with corrections 1972
Reprinted 1979
First paperback edition 1988
Reprinted 1992
Re-issued in this digitally printed version 2008

A catalogue record for this publication is available from the British Library

Library of Congress Catalogue Card Number: 68-18343

ISBN 978-0-521-04722-7 hardback
ISBN 978-0-521-35732-6 paperback

CONTENTS

PREFACE

This book, based on lectures given in the University of St Andrews, is intended to give honours students the background and training necessary before they start to study functional analysis.

There are many books on functional analysis; and some of them seem to go over the preliminaries to the subject far too quickly. The aim here is to provide a more leisurely approach to the theory of the topology of metric spaces, a subject which is not only the basis of functional analysis but also unifies many branches of classical analysis. The applications of the theory in Chapter 8 to problems in classical algebra and analysis show how much can be done without ever defining a normed vector space, a Banach space or a Hilbert space.

The reader is not expected to know much more classical analysis than is contained in Hardy's *Pure Mathematics* or Burkill's *First Course in Mathematical Analysis*. A knowledge of the elements of the theory of uniform convergence is assumed. Analytic functions and Lebesgüe integrals are mentioned occasionally; their introduction provides more advanced applications of the theory of metric spaces, but adds nothing to the theory.

I am most grateful to Professor Arthur Erdélyi and to the Editors of the series of Cambridge Mathematical Tracts for their kindly criticisms and suggestions.

E.T.C.

31 March 1967

I wish to thank Mr Frank Gerrish and Professor Edwin Hewitt for many corrections, which have proved invaluable to me in the preparation of the second impression of this book.

E.T.C.

11 December 1971

CHAPTER 1

INTRODUCTION

1. Sequences and limits

A *sequence* $\{x_n\}$ is a collection of objects occurring in order; thus there is a first member x_1, a second member x_2, and so on indefinitely. For every positive integer k, there is a corresponding kth member of the sequence. The members of such a sequence need not be all different. We can have a sequence all of whose members are the same; such a sequence is called a constant sequence.

If $\{k_n\}$ is a strictly increasing sequence of positive integers, the sequence $\{x_{k_n}\}$ is called a *subsequence* of $\{x_n\}$. The definition implies that $\{x_n\}$ is a subsequence of itself.

A sequence $\{x_n\}$ of real numbers is said to converge to the limit x if, for every positive value of ϵ, all but a finite number of members of the sequence lie between $x-\epsilon$ and $x+\epsilon$. If a sequence $\{x_n\}$ of real numbers converges, every subsequence converges to the same limit. A sequence of real numbers which converges to zero is called a *null-sequence*. Thus if $\{x_n\}$ converges to x, the sequence $\{x_n - x\}$ is a null-sequence.

It is often convenient to represent real numbers by points on a line, and to speak of the point of abscissa x simply as the point x. The distance between the points x and y is $|x-y|$. To say that the sequence of real numbers $\{x_n\}$ converges to x is thus the same thing as saying that the sequence of points $\{x_n\}$ converges to the point x, or that the distance between the point x_n and the point x tends to zero as $n \to \infty$.

In the same way, we frequently speak of the complex number z as the point z in the complex plane. That the sequence $\{z_n\}$ of complex numbers converges to z means that the distance $|z_n - z|$ between the point z_n and the point z in the complex plane tends to zero as $n \to \infty$. This geometrical language is very convenient, but in fact very little geometry is used. All we need in analysis is that the distance between two points satisfies the triangle inequality $|z_1 - z_2| \leqslant |z_1 - z_3| + |z_3 - z_2|$.

As the complex plane is merely a geometrical picture, we are not compelled to use the Euclidean formula $|z_1 - z_2|$ for the distance between two points. In the extended complex plane, that is, the ordinary complex plane with an added ideal point at infinity, it is sometimes more convenient to use as the 'distance' between the points z_1 and z_2 the expression

$$\frac{2|z_1 - z_2|}{\sqrt{\{(1+|z_1|^2)(1+|z_2|^2)\}}}.$$

This is known as the chordal distance since it is the length of the chord joining the points corresponding to z_1 and z_2 on the Riemann sphere whose stereographic projection is the extended complex plane. This 'distance' satisfies the triangle inequality; and when we are dealing with convergence to a finite limit, it does not matter which definition of 'distance' we use.

There are, however, important differences. The complex plane is unbounded, that is, we can find pairs of points whose distances apart are as large as we like. But the extended complex plane with the chordal definition of 'distance' is bounded, since no pair of points is at a 'distance' apart which exceeds 2. The 'distance' from z to the point at infinity is $2/\sqrt{(1+|z|^2)}$. To say that the sequence $\{z_n\}$ tends to the limit ∞ is equivalent to saying that, for every positive number ϵ (< 2), all but a finite number of members of the sequence are at a 'distance' less than ϵ from the point at infinity; it is readily seen that this condition reduces to

$$|z_n| > \left\{\frac{4}{\epsilon^2} - 1\right\}^{\frac{1}{2}}$$

for all but a finite number of values of n, as it should since we can choose ϵ (< 2) so that $(4\epsilon^{-2} - 1)^{\frac{1}{2}}$ is as large as we please.

A sequence might be a sequence $\{x_n(t)\}$ of real functions, each continuous on an interval $a \leqslant t \leqslant b$. If this sequence converged for each value of t in the interval, its limit $x(t)$ would be a function defined on the same interval; $\{x_n(t)\}$ is then said to be pointwise convergent. The sequence is said to converge uniformly to $x(t)$ on the interval if, for every positive value of ϵ, there exists an integer n_0, depending on ϵ but not on t, such that, whenever $n \geqslant n_0$, the inequality

$$|x_n(t) - x(t)| < \epsilon$$

holds at all points of the interval; if this condition is satisfied, the limit-function $x(t)$ is itself continuous.

There is another definition of uniform convergence equivalent to this which depends on the idea of supremum (least upper bound). Suppose that, for every positive integer n, the function $|x_n(t)-x(t)|$ has supremum M_n. This means that

$$|x_n(t)-x(t)| \leqslant M_n \quad (a \leqslant t \leqslant b),$$

and that, for every positive value of ϵ, there exists a number t_0 of the interval such that

$$|x_n(t_0)-x(t_0)| > M_n - \epsilon.$$

The sequence $\{x_n(t)\}$ converges uniformly to $x(t)$ if and only if M_n tends to zero as $n \to \infty$.

This alternative definition of uniform convergence enables us to express the idea in a geometrical form. We call each function $x(t)$, continuous on $a \leqslant t \leqslant b$, a 'point'; and we define the 'distance' between two 'points' $x(t)$ and $y(t)$ as the supremum of $|x(t)-y(t)|$, a definition which satisfies the triangle inequality. To say that the sequence of continuous functions $\{x_n(t)\}$ converges uniformly to the continuous function $x(t)$ is equivalent to saying that the 'distance' from the 'point' $x_n(t)$ to the 'point' $x(t)$ tends to zero as $n \to \infty$.

We might have defined the 'distance' between the 'points' $x(t)$ and $y(t)$ as

$$\left\{\int_a^b |x(t)-y(t)|^2\,dt\right\}^{\frac{1}{2}}.$$

When the sequence of 'points' $\{x_n(t)\}$ converges to $x(t)$ with this definition of 'distance', the sequence of functions $\{x_n(t)\}$ is said to *converge in mean of order* 2 to $x(t)$. Uniform convergence evidently implies convergence in mean of order 2, but not conversely. For example, $\{nt/(1+n^2t^2)\}$ converges in mean of order 2 on $0 \leqslant t \leqslant 1$ to zero, but is not uniformly convergent.

The study of the properties of sets of 'points' in a 'space' whose only geometrical property is the existence of a 'distance' between each pair of 'points' is called *metric space topology*. It had its beginnings in the theory of linear sets of points, which arose in the 19th century from a discussion of the theory of

Fourier series and of the functions which can be represented by such series. An account of this will be found in the first volume of E. W. Hobson's *Theory of Functions of a Real Variable and the Theory of Fourier's Series* (Cambridge, 1921).

We could develop metric space topology as a piece of abstract mathematics, and never mention its applications. Here we shall consider the subject in relation to its applications and show how the ideas unify branches of analysis which may seem disconnected.

In the rest of this chapter, we introduce ideas which may be well-known to the reader but will be needed in the sequel, before proceeding to the definition of a metric space. A knowledge of the elements of classical analysis is assumed.

2. Sets

In the previous section, we used the word 'set'. It is unnecessary here to discuss the logical difficulties involved in the general notion of a set. By a set, we always mean a well-defined collection of distinct elements, called the members of the set. Sometimes we shall use the word 'class' or 'family' instead of 'set'.

A finite set has only a finite number of members. An infinite set has an infinite number of members. An infinite set of real numbers is not necessarily unbounded; for example, the set of all rational numbers between ± 1 is infinite but is bounded. A set is said to be *empty* or *void* if it has no members. For example, the set of all positive integers less than 10 is finite; the set of all positive integers is infinite; the set of all positive integers less than -2 is empty.

An infinite set is said to be *countable* or *denumerable* if its members can be arranged as a sequence. The set of all rational numbers x such that $0 < x \leqslant 1$ is countable. Consider first the rational numbers p/q, between 0 and 1, each expressed in its lowest terms; for each given q, arrange the numbers p/q in order of increasing p; then arrange the groups so obtained in order of increasing q. We get the sequence

$$\tfrac{1}{1}, \tfrac{1}{2}, \tfrac{1}{3}, \tfrac{2}{3}, \tfrac{1}{4}, \tfrac{3}{4}, \tfrac{1}{5}, \tfrac{2}{5}, \tfrac{3}{5}, \tfrac{4}{5}, \tfrac{1}{6}, \tfrac{5}{6}, \tfrac{1}{7} \ldots$$

which proves the result. But it will be proved later that the set of all real numbers between 0 and 1 is not countable: the real numbers between 0 and 1 cannot be arranged as a sequence.

We said that a set is countable if its members can be arranged as a sequence. But a sequence is not a set, not even if all the members of the sequence are distinct. The reason for this is that a set is a collection of distinct entities, considered merely as a collection; but in a sequence the order is important. For example, from the set of positive integers, we can construct many different sequences; 1, 2, 3, 4, 5 ... and 1, 3, 2, 5, 7, 4, 9, 11, 6, ... are two different sequences containing all the positive integers.

Sets are usually denoted by capital letters, members of a set by small letters. The notation $a \in A$ means that a is a member of the set A; $a \notin A$ that a is not a member of A. The empty set—there is only one empty set—is denoted by $\varnothing$.

A finite set can be defined by writing down explicitly the elements which are its members. Thus $\{1, 3, 5, 7\}$ is a set with four members, the odd natural numbers less than 8. Again $\{0\}$ is the set whose only member is the number zero; it is not the empty set —it has one member.

Certain capital letters are reserved to denote particular sets. Some of these are

N: the set of natural numbers 0, 1, 2, 3, ...
R: the set of real numbers.
R^2: the set of all ordered pairs (x, y) of real numbers.
R^3: the set of all ordered triples (x, y, z) of real numbers.
$C[a, b]$: the set of all functions $x(t)$ continuous on $a \leqslant t \leqslant b$.

Sometimes it is convenient to define a set by stating a property which is possessed by all the members of the set and by no element which does not belong to the set. Thus the set $\{2, 3\}$ might be defined as the set of all real numbers x which satisfy the equation $x^2 - 5x + 6 = 0$; we should then denote the set by $\{x \in R: x^2 - 5x + 6 = 0\}$. Again the set $\{1, 3, 5, 7\}$ could be defined as $\{x \in N: x \equiv 1 \pmod 2,\ x < 8\}$. The set $\{x \in N: x \equiv 0 \pmod 2,\ x \equiv 1 \pmod 2\}$ is the empty set, since there is no natural number which is both even and odd.

3. Sets and subsets

The sets A and B are said to be equal, $A = B$, if every member of A belongs to B and every member of B belongs to A.

The set A is said to be contained in B if every member of A belongs to B. We then write $A \subseteq B$. In this case, we may also say that B contains A, and write $B \supseteq A$; or again that A is a subset of B. If $A = B$, then $A \subseteq B$ and $B \supseteq A$; and conversely.

The set A is said to be properly contained in B if every member of A belongs to B and there is at least one member of B which does not belong to A. We then write $A \subset B$ or $B \supset A$, and say that A is a proper subset of B.

Some authors write $A \subset B$ whenever A is a subset of B, not necessarily a proper subset of B, and do not use the notation $\subseteq, \supseteq$.

Evidently,

if $A \subseteq B$ and $B \subseteq C$, then $A \subseteq C$;

if $A \subseteq B$ and $B \subset C$, then $A \subset C$;

if $A \subset B$ and $B \subseteq C$, then $A \subset C$;

if $A \subset B$ and $B \subset C$, then $A \subset C$.

The empty set $\varnothing$ is a subset of every set. For if not, there would exist a set A of which $\varnothing$ is not a subset; this would mean that there is a member of $\varnothing$ which does not belong to A. This is impossible, since $\varnothing$ has no members.

4. The complement of a set

Let us now consider sets which are all subsets of some fixed set E. The complement of a subset A with respect to E is the set of members of E which do not belong to A. We shall denote it by A'. The following results are evident from the definition:

(i) $E' = \varnothing$;

(ii) $\varnothing' = E$;

(iii) $(A')' = A$;

(iv) $A \subseteq B$ if and only if $B' \subseteq A'$.

This notation is satisfactory so long as we are considering only complements with respect to one fixed set E. When complements with respect to several sets occur, other notations are sometimes used, such as $\complement_E A$, or $E-A$.

5. Finite unions and intersections

The union of a family of subsets of the fixed set E is defined to be the set of members of E, each of which belongs to at least one subset of the family. The intersection of a family of subsets is defined to be the set of members of E, each of which belongs to all the subsets of the family. In either case, the family may consist of a finite or of a infinite number of subsets of E; and if it consists of an infinite number of subsets, the family is not necessarily countable.

The *union* $A \cup B$ of two subsets A and B of the fixed set E is the subset of E which consists of those members of E which belong to A or to B or to both.

The *intersection* $A \cap B$ is the subset of E which consists of those members of E which belong to A and to B. If $A \cap B = \varnothing$, A and B have no common member and are said to be *disjoint*.

The intersection $A \cap B'$ of A and the complement of B with respect to E evidently consists of the members of A which do not belong to B. It is called the *relative complement* of B with respect to A.

If A, B, C are three subsets of E, then $(A \cup B) \cup C = A \cup (B \cup C)$.

For if $x \in (A \cup B) \cup C$, then $x \in A \cup B$, or $x \in C$, or x belongs to both. If $x \in C$, $x \in B \cup C$ and therefore $x \in A \cup (B \cup C)$. If $x \in A \cup B$, then x belongs to A or to B or to both; if $x \in A$, $x \in A \cup (B \cup C)$; if $x \in B$, then $x \in B \cup C$ and therefore $x \in A \cup (B \cup C)$. Thus if

$$x \in (A \cup B) \cup C, \quad \text{then} \quad x \in A \cup (B \cup C),$$

and so

$$(A \cup B) \cup C \subseteq A \cup (B \cup C).$$

Similarly $A \cup (B \cup C) \subseteq (A \cup B) \cup C$. Hence the result. We may therefore leave out the brackets and write $A \cup B \cup C$ for this triple union. And, evidently, the triple union does not depend on the order of the sets A, B, C.

If n is any positive integer and $A_1, A_2, \ldots, A_n$ are subsets of E, the set

$$A_1 \cup A_2 \cup A_3 \cup \ldots \cup A_n$$

consists of the members of E which belong to at least one of the subsets A_i. It is denoted by

$$\bigcup_{i=1}^{n} A_i \quad \text{or by} \quad \bigcup_{i} A_i.$$

If A, B, C are three subsets of E, then $(A \cap B) \cap C = A \cap (B \cap C)$. The proof is similar to that given above. Again we may leave out the brackets, and write simply $A \cap B \cap C$. The intersection

$$\bigcap_{i=1}^{n} A_i = A_1 \cap A_2 \cap A_3 \cap \ldots A_n$$

consists of those members of E which belong to all the subsets $A_1, A_2, \ldots, A_n$.

If A and B are any two subsets of E, then $(A \cup B)' = A' \cap B'$ and $(A \cap B)' = A' \cup B'$.

If $x \in (A \cup B)'$, $x \notin A \cup B$. Hence $x \notin A$ and therefore $x \in A'$. Similarly $x \in B'$, and so $x \in A' \cap B'$. Thus $(A \cup B)' \subseteq A' \cap B'$. Again, if $x \in A' \cap B'$, $x \in A'$ and so $x \notin A$; similarly $x \notin B$. Hence $x \notin A \cup B$, that is, $x \in (A \cup B)'$. Thus $A' \cap B' \subseteq (A \cup B)'$. The result follows.

The second result can be proved in the same way. Alternatively, it can be proved by taking complements. Remembering that $A'' = A$, we have $A \cup B = (A \cup B)'' = (A' \cap B')'$. Writing $A' = C$, $B' = D$, we obtain $A = A'' = C'$, $B = B'' = D'$; and so $C' \cup D' = (C \cap D)'$, as required.

6. Infinite unions and intersections

Let S be any set, finite or infinite; it might be the positive integers between 0 and 10, it might be all the positive integers, it might even be all the real numbers between 0 and 1. Suppose that, corresponding to each member of α of S, there exists a subset of E which we denote by A_α. We then have a family of subsets of E with index set S. We denote the union and intersection of the family by

$$\bigcup_{\alpha \in S} A_\alpha, \quad \bigcap_{\alpha \in S} A_\alpha.$$

Evidently

$$(\bigcup_{\alpha \in S} A_\alpha)' = \bigcap_{\alpha \in S} A'_\alpha, \quad (\bigcap_{\alpha \in S} A_\alpha)' = \bigcup_{\alpha \in S} A'_\alpha.$$

The union of a countable family of countable sets is countable.

A countable family of sets can be arranged as a sequence $A_1, A_2, A_3, \ldots$. Let $A_m = \{a_{m,1}, a_{m,2}, a_{m,3}, \ldots\}$. Then $\bigcup_{m=1}^{\infty} A_m$ consists of all elements $a_{m,n}$, and these can be arranged as a sequence

$$a_{1,1}, a_{1,2}, a_{2,1}, a_{1,3}, a_{2,2}, a_{3,1}, a_{1,4}, \ldots.$$

If the sets are not disjoint, there will be repetitions in this sequence. But the elements of $\bigcup_{m=1}^{\infty} A_m$ form a subsequence of this sequence when repetitions are omitted. Hence $\bigcup_{m=1}^{\infty} A_m$ is countable.

It follows from this that the set of all rational numbers is countable. We have seen that the set A_0 consisting of all rational numbers x, such that $0 < x \leqslant 1$ is countable. It follows that the set A_1, consisting of all rational numbers x such that $1 < x \leqslant 2$ is countable since they are in one-one correspondence with the members of A_0. Similarly for A_n, the set of all rational numbers x such that $n < x \leqslant n+1$ for any positive or negative integer n. The set of all rational numbers is

$$A_0 \cup A_1 \cup A_{-1} \cup A_2 \cup A_{-2} \cup \ldots;$$

hence the result.

7. The algebra of sets

Let A, B, C be subsets of a given set E. Then the operations of union, intersection and complementation have the following properties:

(*a*) $A \cup A' = E,\qquad A \cap A' = \varnothing.$

(*b*) $A \cup A = A,\qquad A \cap A = A.$

(*c*) $A \cup \varnothing = A,\qquad A \cap E = A.$

(*d*) $A \cap \varnothing = \varnothing,\qquad A \cup E = E.$

(*e*) $A \cup B = B \cup A,\qquad A \cap B = B \cap A.$

(*f*) $(A \cup B) \cup C = A \cup (B \cup C),\qquad (A \cap B) \cap C = A \cap (B \cap C).$

(*g*) $A \subseteq A \cup B,\qquad A \supseteq A \cap B.$

(*h*) $(A \cup B) \cap C = (A \cap C) \cup (B \cap C),$

$$(A \cap B) \cup C = (A \cup C) \cap (B \cup C).$$

These properties have been arranged in pairs; each relation in a pair can be deduced from the other by using the duality property of the previous section, viz. $(A \cup B)' = A' \cap B'$ and its equivalent $(A \cap B)' = A' \cup B'$. Thus it is necessary only to prove half these formulae; in fact, many of them are obvious from the definitions of union and intersection.

In (g), it is evident that $A \subseteq A \cup B$. Hence

$$A' \supseteq (A \cup B)' = A' \cap B'.$$

Replace A', B' by C, D. Then $C \supseteq C \cap D$.

To prove the first formula in (h), note that, if $x \in (A \cup B) \cap C$, then x belongs to $A \cup B$ and to C. But if $x \in A \cup B$, x belongs to A or to B or to both; hence x belongs to $A \cap C$ or to $B \cap C$ or to both. Thus $x \in (A \cap C) \cup (B \cap C)$, and so

$$(A \cup B) \cap C \subseteq (A \cap C) \cup (B \cap C).$$

Again, if $x \in (A \cap C) \cup (B \cap C)$, x belongs to $A \cap C$ or to $B \cap C$ or to both. If $x \in A \cap C$, x belongs to A and to C; therefore $x \in A \cup B$ and $x \in C$ and so $x \in (A \cup B) \cap C$; and similarly if $x \in B \cap C$. Hence $(A \cap C) \cup (B \cap C) \subseteq (A \cup B) \cap C$, which completes the proof.

To obtain the second formula in (h), we take the complement of the first formula, which gives

$$(A \cup B)' \cup C' = (A \cap C')\cup(B \cap C)'.$$

Hence $$(A' \cap B') \cup C' = (A' \cup C') \cap (B' \cup C')$$

or, replacing A', B', C' by their complements,

$$(A \cap B) \cup C = (A \cup C) \cap (B \cup C).$$

8. Equivalence relations

Two natural numbers a and b may be connected by one of many relations; for example, $a < b$, $a \equiv b \pmod 3$, a is a divisor of b, a is a multiple of b, and so on. All these are examples of what are called *binary relations*. Such relations occur in many other branches of mathematics. In plane geometry, the property that a point a is conjugate to a point b with respect to a fixed conic is a binary relation. They also occur in every-day life: on the set of all human beings, we have, for example the binary relations 'a is the father of b' or 'a is of the same nationality as b', and so on. If we have a given relation on a set E, we often write $a \sim b$ to indicate that the element a of E stands in the given relationship to the element b of E.

Whilst it is obvious what a relation is, it is rather difficult to define. One way of defining a relation involves introducing the *Cartesian product* of two sets. Let X and Y be sets. By taking first a member x of X and then a member y of Y, we obtain an ordered pair (x, y). The set of all such ordered pairs is called the Cartesian product of X and Y, and is denoted by $X \times Y$.

Now suppose we have a relation $\sim$ defined *on* a set E. This means that for each x belonging to E, there is at least one y belonging to E such that $x \sim y$; in other words, for each x which belongs to E, there are one or more ordered pairs (x, y) such that y belongs to E and $x \sim y$. But the set of all such ordered pairs is a non-empty subset of the Cartesian product set $E \times E$. Thus a binary relation on E is a non-empty subset of $E \times E$ of a special type.

Let $\sim$ be a binary relation defined *on* a set E. It is said to be *reflexive* if $a \sim a$ for all a belonging to E. It is said to be *symmetric* if $a \sim b$ implies $b \sim a$. It is said to be *transitive* if $a \sim b$, $b \sim c$ implies $a \sim c$. For example, on the set N of natural numbers, the relation $<$ is transitive but not reflexive or symmetric; the relation 'is a divisor of' is reflexive and transitive, but it is not symmetric; the relation of 'congruence (mod 3)' is reflexive, symmetric and transitive. A binary relation which is reflexive, symmetric, and transitive is called an *equivalence relation.*

If $\sim$ is an equivalence relation on E, the set of all members x of E such that $x \sim a$ is called an equivalence class; let us denote it temporarily by $\bar{a}$. For example, consider the equivalence relation of congruence (mod 3) on the set of natural numbers; there are three equivalence classes

$$\{0, 3, 6, 9, \ldots\}, \quad \{1, 4, 7, 10, \ldots\} \quad \{2, 5, 8, 11, \ldots\}.$$

If $\sim$ is an equivalence relation on a set E, $a \sim b$ implies $\bar{a} = \bar{b}$, and conversely.

Suppose that $a \sim b$. Let $c \in \bar{a}$. Then $c \sim a$. Since the relation is transitive, this implies that $c \sim b$ so that $c \in \bar{b}$. Hence $\bar{a} \subseteq \bar{b}$. But $a \sim b$ implies $b \sim a$ since the relation is symmetric, and hence $\bar{b} \subseteq \bar{a}$. Therefore $\bar{a} = \bar{b}$.

Now suppose that $\bar{a} = \bar{b}$. By the reflexive law, $b \sim b$ and so $b \in \bar{b}$. But this implies that $b \in \bar{a}$, that is, $b \sim a$.

By a *partition* of E, we mean a collection of non-empty subsets of E such that each member of E belongs to one and only one subset of the collection.

If $\sim$ is an equivalence relation on a set E, two equivalence classes are either identical or have no common member; the collection of all equivalence classes is a partition of E.

If $\bar{a}$ and $\bar{b}$ have a common member c, then $c \sim a$ and $c \sim b$. By the symmetric and transitive laws, $a \sim b$, and hence $\bar{a} = \bar{b}$. Thus if $\bar{a}$ and $\bar{b}$ are different, they have no common member.

Let d be any member of E. Since $d \sim d$, $d \in \bar{d}$. Hence every member of E belongs to an equivalence class; and since two equivalence classes cannot overlap, it belongs to precisely one equivalence class.

9. The axioms of the system of real numbers

There are various ways of defining the real number system. One can define a real number as a decimal form which may terminate, recur or not recur, as is done, for example in P. Dienes's book *The Taylor Series* (Oxford, 1931). One can define a real number as a Dedekind section of the rational numbers. One can define a real number as an equivalence class of equivalent Cauchy sequences of rational numbers, the Cantor construction. The interesting thing is that it does not matter which definition we adopt. Each method generates an Archimedean ordered field with the supremum property, and any two such fields are isomorphic. In popular language, this means that, whatever definition of the real number system is adopted, any arithmetical operation leads to the same answer.

A *field* is defined to be a set S of elements which may be combined by two operations, called addition and multiplication; and these operations satisfy the following basic laws.

(i) *The laws of closure.* Addition and multiplication are well defined. This means that, for every ordered pair a and b of members of S, there are unique elements $a+b$ and $a \times b$ belonging to S.

(ii) *The commutative laws.* For every pair a and b of members of S,

$$a+b = b+a, \quad a \times b = b \times a.$$

(iii) *The associative laws.* For any three members a, b and c of S,

$$(a+b)+c = a+(b+c), \quad (a \times b) \times c = a \times (b \times c).$$

(iv) *The existence of zero and unity.* S contains unique and distinct elements 0 and 1 such that, for every member a of S,

$$a+0 = a, \quad a \times 1 = a.$$

(v) *The existence of additive and multiplicative inverses.* For every a belonging to S, there exists a unique member of S, denoted by $-a$, such that $a+(-a) = 0$. For every $a \neq 0$ belonging to S, there exists a unique member of S, denoted by a^{-1}, such that $a \times a^{-1} = 1$.

(vi) *The distributive law*

$$a \times (b+c) = (a \times b)+(a \times c).$$

Briefly, a field is a commutative additive group; if we omit the element 0, it is also a commutative multiplicative group; and it satisfies the distributive law (vi).

A field S is said to be *ordered* if it contains a set P of 'positive' members with the properties:

(i) P is closed with respect to addition and multiplication.

(ii) If a is an element of the field S, then either $a = 0$, or $a \in P$ or $-a \in P$.

When this condition is satisfied, the relations $a < b$ and $b > a$ both mean that $b-a$ (defined to be $b+(-a)$) belongs to P. The set of rational numbers, with 0 and 1 having their ordinary meanings, is an ordered field.

An ordered field S is said to have the supremum property if and only if every non-empty subset of S, which is bounded above, has a supremum in S.

A subset A of S is said to be bounded above if there is an element k in S such that $x < k$ for every member x of A. Such an element is called an upper bound. If k is an upper bound, any element of S greater than k is an upper bound. The point of the definition is that the set of upper bounds of A has a least member, which is called the *supremum* or least upper bound of A, and which is denoted by $\sup A$ or $\sup_{x \in A} x$ or $\sup \{x : x \in A\}$. If $k_1 = \sup A$, then k_1 and every member of S greater than k_1 is an upper bound, but no member of S less than k_1 is an upper bound. An equivalent form of the definition is:

An ordered field S has the supremum property if and only if every non-empty subset of S which is bounded below has an infimum in S.

The *infimum* of a subset A is its greatest lower bound, and is denoted by $\inf A$ or $\inf_{x \in A} x$ or $\inf \{x : x \in A\}$.

The ordered field Q of rational numbers does not have the supremum property. For example, the subset A of Q, consisting of the partial sums of the series

$$1 + \frac{1}{1!} + \frac{1}{2!} + \frac{1}{3!} + \ldots,$$

is bounded above, since each partial sum is less than 3. But there is no rational number which is the supremum of A.

These are the axioms of a real number system. A detailed discussion will be found for example in Birkhoff and MacLane's *Survey of Modern Algebra* (New York, 1953).

10. Properties of the real numbers

By a system of real numbers, we mean an ordered field with the supremum property. In analysis, we can use any of the representations to which we have referred; it does not matter which. We often do not need to use a representation; we only need certain basic properties which follow from the axioms and do not depend on a concrete representation.

Let R be a system of real numbers. The positive integers in R are defined in the obvious way, by $1+1 = 2$, $1+2 = 3$, and so on. The negative integers and their properties follow from the properties of the additive inverse. If p and q are integers and q is not zero, the rational number $p \div q$ is defined to be $p \times q^{-1}$, where q^{-1} is the multiplicative inverse of q. All the elementary arithmetic of rational numbers follows from the field properties.

If a and b are two positive real numbers, there exists a positive integer k such that $ka > b$.

This is the Archimedean property of the system R of real numbers. If it were not true, there would exist a pair of positive real numbers a and b such that $na \leqslant b$ for every positive integer n. The sequence $\{na\}$ thus has b as an upper bound, and, by the supremum property, it has a supremum c; hence $na \leqslant c$ for every positive integer n. Therefore $(n+1)a \leqslant c$ for every positive integer n. But this gives $na \leqslant c-a$. Thus $c-a$ is an upper bound of the sequence $\{na\}$, which is impossible since a is positive and c is the least upper bound. This is a contradiction. Hence R does have the Archimedean property.

Between any two real numbers a and b, where $a > b$, there exists a rational number.

By hypothesis, $a-b > 0$. By the Archimedean property, there exists an integer n such that $n(a-b) > 1$, that is, $a-b > 1/n$. With this n, there are integers k such that $k > nb$; let m be the least k which has this property. Then $m-1 \leqslant nb$. Hence

$$b < \frac{m}{n} = \frac{m-1}{n} + \frac{1}{n} < b + (a-b) = a$$

so that

$$b < \frac{m}{n} < a.$$

Every real number is the supremum of a set of rational numbers.

If a is any real number, let S be the set of all rational numbers less than or equal to a. The set S is bounded above by a; let b be the supremum of S. Evidently $b \leqslant a$. If $b < a$, there exists a rational number between b and a, which is impossible since b is the supremum of all rational numbers less than or equal to a. Hence $b = a$, which was to be proved.

There is no sequence which contains all the real numbers.

It suffices to prove that there is no sequence which contains all the real numbers between 0 and 1. We assume that every such number can be represented as a non-terminating decimal fraction, the terminating decimals being regarded as recurring decimals ending with 0. The representation is unique if we disallow decimals involving 9.

If the real numbers between 0 and 1 formed a countable set, they could be arranged as a sequence $\{x_n\}$. The real number x_n would have decimal representation

$$0{\cdot}x_{n,1}x_{n,2}x_{n,3}\ldots,$$

where each $x_{n,r}$ is one of the integers 0, 1, 2, ..., 9. Now let

$$y_r = 1, 2, 3, 4, 5, 6, 7, 8, 7 \text{ or } 0$$

according as

$$x_{r,r} = 0, 1, 2, 3, 4, 5, 6, 7, 8 \text{ or } 9.$$

Then

$$y = 0{\cdot}y_1 y_2 y_3 \ldots$$

is a decimal fraction different from all the members of the sequence $\{x_n\}$ and it lies between 0 and 1. This is impossible since the sequence by hypothesis contained all real numbers between 0 and 1. Hence the real numbers between 0 and 1 do not form a countable set.

11. Sequences of real numbers

Every bounded monotonic sequence of real numbers is convergent.

Let $\{a_n\}$ be a bounded increasing sequence of real numbers. Let $a = \sup\{a_n\}$, so that a is a well-defined finite real number. For every positive value of ϵ, a is an upper bound of the sequence, but $a-\epsilon$ is not. Hence $a_n \leqslant a$ for all values of n, but $a_m > a-\epsilon$ for at least one positive integer m. Therefore, whenever $n \geqslant m$,

$$a-\epsilon < a_m \leqslant a_n \leqslant a,$$

which proves the result. Similarly if $\{a_n\}$ is a bounded decreasing sequence.

Every bounded sequence of real numbers contains a convergent subsequence.

If $\{a_n\}$ is a bounded sequence of real numbers, consider the set

S of real numbers x such that $a_n > x$ for at most a finite number of values of n. The set S is bounded below since $\{a_n\}$ is a bounded sequence; let H be the infimum of S. We show that there is a subsequence of $\{a_n\}$ which converges to H.

For every positive value of ϵ, $a_n > H+\epsilon$ for at most a finite number of values of n, but $a_n > H-\epsilon$ for an infinite number of values of n. Let k_1 be the least positive integer such that a_{k_1} lies between $H \pm 1$. Let k_2 be the least positive integer greater than k_1 such that a_{k_2} lies between $H \pm \frac{1}{2}$. Next let k_3 be the least positive integer greater than k_2 such that a_{k_3} lies between $H \pm \frac{1}{3}$. And so on indefinitely, by the definition of H. In this way, we construct a strictly increasing sequence $\{k_n\}$ of positive integers such that, for every positive integer n,

$$H-\frac{1}{n} < a_{k_n} < H+\frac{1}{n}.$$

Hence the subsequence $\{a_{k_n}\}$ converges to H.

No subsequence can converge to a limit greater than H. The number H, which is denoted by $\limsup_{n\to\infty} a_n$ or by $\overline{\lim}_{n\to\infty} a_n$, is sometimes called the *limes superior* of the sequence $\{a_n\}$.

By a similar argument, we can prove the existence of a real number h with the property that no subsequence can converge to a limit less than h and that there exists a subsequence which converges to h. The number h, which is denoted by $\liminf_{n\to\infty} a_n$ or by $\underline{\lim}_{n\to\infty} a_n$, is sometimes called the *limes inferior* of the sequence.

If the sequence $\{a_n\}$ converges to a limit a, we have $H = h = a$, since every subsequence converges to a. Conversely if $H = h$, the sequence is convergent.

A sequence of real numbers is said to be a *Cauchy sequence* if, for every positive value of ϵ, there exists a positive integer n_0 such that

$$|a_m - a_n| < \epsilon$$

whenever $m > n \geqslant n_0$.

A necessary and sufficient condition for a sequence to be convergent is that it be a Cauchy sequence.

This result is known as *Cauchy's principle of convergence.*

Let $\{a_n\}$ be a sequence of real numbers which converges to the real number a. Then, for every positive value of ϵ, there exists a positive integer n_0 such that

$$|a_n - a| < \tfrac{1}{2}\epsilon$$

whenever $n \geqslant n_0$. It follows that, if $m > n \geqslant n_0$,

$$|a_m - a_n| < \epsilon.$$

The sequence is therefore a Cauchy sequence.

Conversely, let $\{a_n\}$ be a Cauchy sequence, so that, for every positive value of ϵ, there exists a positive integer n_0 such that

$$|a_m - a_n| < \epsilon,$$

whenever $m > n \geqslant n_0$. This implies that $|a_m| < |a_{n_0}| + \epsilon$, so that the sequence is bounded. It follows that the sequence contains a subsequence $\{a_{k_n}\}$ which converges to a limit a. Since $k_m \geqslant m$, we have

$$|a_{k_m} - a_n| < \epsilon,$$

whenever $m > n \geqslant n_0$. Make $m \to \infty$; then

$$|a - a_n| \leqslant \epsilon,$$

whenever $n \geqslant n_0$, which means that the sequence $\{a_n\}$ converges to a. The condition is thus sufficient.

In the terminology to be introduced in Chapter IV, the real number system is *complete*.

12. Functions and mappings

In the theory of functions of a complex variable, we meet the idea of a conformal mapping. We say, for example, that the half-plane $\operatorname{re} z > 0$ is mapped conformally onto the unit disc $|w| < 1$ by the relation

$$w = \frac{z-1}{z+1}.$$

When we say that this is a mapping, we mean that the relation associates with each point z of the half-plane a single well-defined point w of the disc. (That this mapping conserves angle is of no interest in the present context). This idea of a mapping can be extended easily to abstract sets.

Let E and F be two non-empty sets (possibly even the same set). By a mapping f of E *into* F, we mean a relation which associates with each element a of E a single well-defined element of F, which is denoted by $f(a)$ and called the image of a under f. This mapping is denoted by $f:E \to F$. The set E is called the *domain* of the mapping. The set of all points of F which are images of points of E is called the *range* of the mapping. The range may be the whole of F or a proper subset of F; it may even be a single point of F, in which case the mapping is called a constant mapping. In the example drawn from complex function theory, the domain was the half-plane $\text{re}\, z > 0$, the range the disc $|w| < 1$, since each point of the disc is the image of a point of the half-plane.

In elementary calculus, we often use a different representation of a function—its graph. The graph of $y = x^2$ picks out all the points (x, y) of the Cartesian plane for which the ordinate y is the square of the abscissa x—it picks out all the ordered pairs (x, y) for which y stands in the given relation to x. The graph of the function is thus a subset of the Cartesian plane.

Given two abstract sets E and F, the Cartesian product $E \times F$ was defined as the set of all ordered pairs (x, y) where $x \in E$, $y \in F$. A function or mapping $f: E \to F$ can be defined as a subset f of $E \times F$ with the property that, for each $a \in E$, f contains exactly one element $(a, f(a))$.

This definition of function or mapping is not restricted to be merely a generalization of a function of one variable. The domain of the mapping could quite well be a Cartesian product set $X \times Y$. The mapping $f: X \times Y \to F$ then associates with each ordered pair (x, y) where $x \in X$, $y \in Y$ a single well-defined element of F.

In the theory of functions of a complex variable, the sum $f(z)$ of a power series $\sum_0^\infty a_n z^n$ is an analytic function regular on its disc of convergence $|z| < R$. If $0 < R < \infty$, it is sometimes possible to continue $f(z)$ analytically outside the disc. By this we mean that it is possible to find an analytic function $g(z)$ regular in a region of which $|z| < R$ is a proper subset; and on the disc of convergence, $g(z) = f(z)$. The process of analytical continuation extends the domain of definition of the analytic function $f(z)$.

A similar idea may arise in the mapping of one abstract set into another. Let E and F be abstract sets, and let E_1 be a proper subset of E. Suppose that $f: E_1 \to F$ is a mapping of E_1 into F. If there is a mapping $g: E \to F$ such that $f(x) = g(x)$ for every x belonging to E_1, g is called an *extension* of f, and f the *restriction* of g to E_1.

CHAPTER 2

METRIC SPACES

13. The definition of a metric space

Let E be any set. Let $\rho(x, y)$ be a function defined on the set $E \times E$ of all ordered pairs (x, y) of members of E, and satisfying the following conditions:

(i) $\rho(x, y)$ is a finite real number for every pair (x, y) of $E \times E$;

(ii) $\rho(x, y) = 0$ if and only if $x = y$;

(iii) $\rho(y, z) \leqslant \rho(x, y) + \rho(x, z)$, where x, y, z are any three elements of E.

Such a function $\rho(x, y)$ is called a *metric* on E; it is a mapping of $E \times E$ into R. A set E with metric ρ is called a *metric space*. The members of E are frequently called 'points', and the function $\rho(x, y)$ the 'distance' from the 'point' x to the 'point' y.

Different choices of metric on a given set E give rise to different metric spaces. For example, the metric space consisting of all complex numbers with metric $|x-y|$ is not the same as that consisting of all complex numbers with metric

$$\frac{2|x-y|}{\sqrt{\{1+|x|^2\}}\sqrt{\{1+|y|^2\}}};$$

and both are different from that consisting of all complex numbers with metric

$$\frac{|x-y|}{1+|x-y|}.$$

14. Properties of the metric

If, in (iii), we put $z = y$, we get

$$2\rho(x, y) \geqslant \rho(y, y) = 0$$

so that $\rho(x, y)$ is a *non-negative* real function. If we put $z = x$, we obtain

$$\rho(y, x) \leqslant \rho(x, y) + \rho(x, x) = \rho(x, y).$$

But as x and y are arbitrary points, a similar argument gives

$$\rho(x,y) \leqslant \rho(y,x).$$

Hence $\rho(x,y) = \rho(y,x)$; thus $\rho(x,y)$ is a *symmetric* function.

Using the symmetry property, we obtain from (iii)

$$\rho(y,z) \leqslant \rho(y,x)+\rho(x,z)$$

or changing the symbols,

(iv) $$\rho(x,y) \leqslant \rho(x,z)+\rho(z,y).$$

If x, y, z are points in the complex plane with the usual Euclidean metric, this inequality asserts that the sum of the lengths of two sides of a triangle is not less than the length of the third side. For this reason, (iii) or its equivalent (iv) is called the *Triangle Inequality*.

Since

$$\rho(x,y)-\rho(x,z) \leqslant \rho(z,y) = \rho(y,z),$$

and

$$\rho(x,z)-\rho(x,y) \leqslant \rho(y,z),$$

we have

(v) $$\rho(y,z) \geqslant |\rho(x,y)-\rho(x,z)|.$$

15. Bounded and unbounded metric spaces

Let M be a metric space consisting of a set E and a metric ρ on E. If there exists a positive number k such that $\rho(x,y) \leqslant k$ for every pair of points x and y of E, we say that M is a *bounded* metric space. A metric space which is not bounded is said to be *unbounded*; in that case, $\rho(x,y)$ takes values as large as we please.

If the metric space M consisting of a set E with metric ρ is unbounded, we can define in many ways a bounded metric space M_1 consisting of the same set E with a different metric. For example, it suffices to put

$$\rho_1(x,y) = \frac{\rho(x,y)}{1+\rho(x,y)}.$$

That this function does satisfy the conditions for a metric is proved as follows.

(i) Since $\rho \geqslant 0$, $\rho_1(x,y) = 0$ if and only if $\rho(x,y) = 0$, that is, if and only if $x = y$.

$$\text{(ii)}\ \rho_1(x,y)+\rho_1(x,z) = \frac{\rho(x,y)}{1+\rho(x,y)}+\frac{\rho(x,z)}{1+\rho(x,z)}$$

$$\geqslant \frac{\rho(x,y)+\rho(x,z)}{1+\rho(x,y)+\rho(x,z)}$$

$$= 1\bigg/\left\{1+\frac{1}{\rho(x,y)+\rho(x,z)}\right\}$$

$$\geqslant 1\bigg/\left\{1+\frac{1}{\rho(y,z)}\right\}$$

$$= \rho_1(y,z).$$

Since $0 \leqslant \rho_1 \leqslant 1$, M_1 is a bounded metric space. The metric ρ_1 is important because it is closely related to the metric ρ. As we shall see in §61, the family of open sets in M is identical with the family of open sets in M_1; the two metrics are then said to be equivalent.

16. The diameter of a subset of a metric space

Let A be a non-empty subset of a metric space M. Consider the set of non-negative real numbers

$$\{\rho(x,y): x \in A,\ y \in A\}.$$

If this set is bounded, it has a supremum, denoted by $\delta(A)$, which is called the diameter of A. If the set is unbounded, it is conventional to write $\delta(A) = +\infty$, and to say that the set has an infinite diameter. It can be easily shown that a necessary and sufficient condition for a non-empty set A to consist of a single point is that $\delta(A) = 0$.

The diameter $\delta(M)$ of the whole space is finite if M is bounded, but is infinite if M is unbounded. For every subset A of M, $\delta(A) \leqslant \delta(M)$.

17. The distance between two subsets of a metric space

Let A and B be two non-empty subsets of a metric space M. The set of real numbers

$$\{\rho(x,y): x \in A,\ y \in B\},$$

is bounded below by zero. Its infimum, denoted by $d(A,B)$,

is called the distance between the subsets A and B. Since $\rho(x,y) = \rho(y,x), d(A,B) = d(B,A)$.

The equation $d(A, B) = 0$ does not imply that A and B have points in common. For example, if A and B are the subsets $x > 0$ and $x < 0$ of the real line with the usual metric, $d(A,B) = 0$, yet A and B have no common point.

If A consists of a single point a,

$$d(A,B) = \inf\{\rho(a,y) : y \in B\}.$$

This is called the distance of the point a from the subset B, and is denoted by $d(a,B)$. The equation $d(a,B) = 0$ does not imply that a belongs to B.

Lastly, if A and B consist of single points a and b respectively, the distance between the subsets A and B reduces to $\rho(a,b)$.

18. Subspaces

Let M be a metric space consisting of a set of points E and a metric ρ; and let E_1 be a proper subset of E. The metric ρ is a function defined on $E \times E$. Let ρ_1 be its restriction to the subset $E_1 \times E_1$, that is,

$$\rho_1(x,y) = \rho(x,y)$$

whenever x and y are points of E_1. Then ρ_1 is a metric on E_1: it is called the *induced metric*. The set E_1 with metric ρ_1 is called a *subspace* of M.

19. Examples of metric spaces

The simplest metric is the discrete metric which can be used to metrize any set E. It is defined as follows: if x and y are any two points of E,

$$\rho(x,x) = 0, \quad \rho(x,y) = 1 \quad (x \neq y).$$

Its properties are very special, but it is sometimes useful. Any space with discrete metric is bounded.

The set R of all real numbers with metric $\rho(x,y) = |x-y|$ is a metric space. It is an unbounded metric space which is called the *real line*.

The extended real number system consists of all real numbers together with two ideal elements $+\infty$ and $-\infty$, with the property that $-\infty < x < \infty$ for every real number x. If we define a function $f(x)$ by

$$f(x) = \frac{x}{1+|x|}, \quad f(+\infty) = 1, \quad f(-\infty) = -1,$$

the function

$$\rho(x,y) = |f(x)-f(y)|$$

is a metric on the extended set of real numbers. The resulting bounded metric space is called the *extended real line*.

This is not the only metric we could apply to the extended real number system. We could apply the discrete metric. Another possible metric is

$$\rho(x,y) = |\tan^{-1} x - \tan^{-1} y|,$$

where $-\frac{1}{2}\pi < \tan^{-1} x < \frac{1}{2}\pi$ when $-\infty < x < \infty$, $\tan^{-1}(+\infty) = \frac{1}{2}\pi$, $\tan^{-1}(-\infty) = -\frac{1}{2}\pi$. Again we get a bounded metric space, since no two points are at a distance which exceeds π.

The metric space C of all complex numbers with metric $|x-y|$ is also an unbounded space; it is called the complex plane or the Argand plane. The extended complex number system consists of all complex numbers together with an ideal number, infinity. One representation of the extended system is provided by the stereographic projection of the sphere $\xi^2+\eta^2+\zeta^2 = 1$ onto the plane $\zeta = 0$, taking the point $(0,0,1)$ as vertex of projection.

Corresponding to the point (ξ, η, ζ) of the sphere we get on $\zeta = 0$ the point of affix

$$z = \frac{\xi + i\eta}{1-\zeta};$$

corresponding to the point of affix z on the plane we get the point (ζ, η, ζ) where

$$\xi + i\eta = \frac{2z}{1+|z|^2}, \quad \zeta = \frac{|z|^2-1}{|z|^2+1};$$

corresponding to the point at infinity, we have the vertex of projection $(0,0,1)$. The chordal distance between the points corresponding to any pair of points of the extended complex plane, viz.

$$\rho(x,y) = \frac{2|x-y|}{\sqrt{(1+|x|^2)}.\sqrt{(1+|y|^2)}}, \quad \rho(x,\infty) = \frac{2}{\sqrt{(1+|x|^2)}},$$

satisfies the condition for a metric. The resulting metric space is denoted by $\bar{C}$.

Euclidean space R^n consists of all ordered n-tuplets

$$\mathbf{x} = (x_1, x_2, \ldots, x_n)$$

of real numbers with metric

$$\rho(\mathbf{x}, \mathbf{y}) = \sqrt{\sum_1^n (x_r - y_r)^2}.$$

The space C^n, consisting of all ordered n-tuplets

$$\mathbf{x} = (x_1, x_2, \ldots, x_n)$$

of complex numbers with metric

$$\rho(\mathbf{x}, \mathbf{y}) = \sqrt{\sum_1^n |x_r - y_r|^2},$$

is merely R^{2n}.

20. The inequalities of Hölder and Minkowski

The graph of $y = \log x$, where $x > 0$, is concave downwards. It follows that, if a and b are positive and if θ (where $0 < \theta < 1$) is fixed, then

$$\theta \log a + (1-\theta)\log b \leqslant \log(\theta a + (1-\theta)b)$$

with equality if and only if $a = b$. This gives

$$a^\theta b^{1-\theta} \leqslant \theta a + (1-\theta)b,$$

a result which evidently also holds so long as a and b are not negative.

In this inequality, put

$$a = a_r/A, \quad b = b_r/B \quad (r = 1, 2, \ldots, n),$$

where a_r and b_r are positive and

$$A = \sum_1^n a_r, \quad B = \sum_1^n b_r,$$

and add. We get

$$\frac{1}{A^\theta B^{1-\theta}} \sum_1^n a_r^\theta b_r^{1-\theta} \leqslant \frac{\theta}{A} \sum_1^n a_r + \frac{1-\theta}{B} \sum_1^n b_r = 1$$

and so
$$\sum_1^n a_r^\theta b_r^{1-\theta} \leqslant \left(\sum_1^n a_r\right)^\theta \left(\sum_1^n b_r\right)^{1-\theta},$$

with equality if and only if

$$\frac{a_r}{A} = \frac{b_r}{B} \quad (r = 1, 2, \ldots, n),$$

that is, if and only if $a_r = kb_r$ for every value of r, k being independent of r.

Now write

$$a_r^{\theta} = c_r, \quad b_r^{1-\theta} = d_r,$$

where c_r and d_r are not negative,

$$\theta = \frac{1}{p}, \quad 1-\theta = \frac{1}{q},$$

where p and q are greater than unity and

$$\frac{1}{p}+\frac{1}{q} = 1.$$

It follows that, if $c_r \geqslant 0$, $d_r \geqslant 0$,

$$\sum_1^n c_r d_r \leqslant \left(\sum_1^n c_r^p\right)^{1/p} \left(\sum_1^n d_r^q\right)^{1/q},$$

with equality if and only if $c_r^p = kd_r^q$ for every value of r, k being independent of r. This result is known as *Hölder's Inequality*. The particular case when $p = 2$ is called *Cauchy's Inequality*.

Now consider

$$\sum_1^n (a_r+b_r)^p,$$

where $a_r \geqslant 0$, $b_r \geqslant 0$ and $p > 1$. Write, as before,

$$\frac{1}{q} = 1-\frac{1}{p}.$$

Then

$$\sum_1^n (a_r+b_r)^p = \sum_1^n a_r(a_r+b_r)^{p-1} + \sum_1^n b_r(a_r+b_r)^{p-1}$$

$$\leqslant \left(\sum_1^n a_r^p\right)^{1/p} \left(\sum_1^n (a_r+b_r)^{q(p-1)}\right)^{1/q}$$

$$+ \left(\sum_1^n b_r^p\right)^{1/p} \left(\sum_1^n (a_r+b_r)^{q(p-1)}\right)^{1/q}.$$

But $q(p-1) = p$, and $1-q^{-1} = p^{-1}$. Hence

$$\left(\sum_1^n (a_r+b_r)^p\right)^{1/p} \leqslant \left(\sum_1^n a_r^p\right)^{1/p} + \left(\sum_1^n b_r^p\right)^{1/p}$$

with equality if and only if $a_r = kb_r$ for every value of r, k being independent of r. This is known as *Minkowski's Inequality*.

In the case $p = 2$, these inequalities reduce to

$$\left(\sum_1^n a_r b_r\right)^2 \leqslant \sum_1^n a_r^2 \sum_1^n b_r^2,$$

$$\left\{\sum_1^n (a_r+b_r)^2\right\}^{\frac{1}{2}} \leqslant \left\{\sum_1^n a_r^2\right\}^{\frac{1}{2}} + \left\{\sum_1^n b_r^2\right\}^{\frac{1}{2}}.$$

The corresponding results for integrals can be proved in the same way. They are that, if $f \geqslant 0$, $g \geqslant 0$, $p > 1$, $q = p/(p-1)$, then

$$\int_\alpha^\beta fg\,dx \leqslant \left\{\int_\alpha^\beta (f(x))^p\,dx\right\}^{1/p} \left\{\int_\alpha^\beta (g(x))^q\,dx\right\}^{1/q},$$

$$\left\{\int_\alpha^\beta (f+g)^p\,dx\right\}^{1/p} \leqslant \left\{\int_\alpha^\beta (f(x))^p\,dx\right\}^{1/p} + \left\{\int_\alpha^\beta (g(x))^p\,dx\right\}^{1/p},$$

whenever the integrals exist. The limits α, β, where $\alpha < \beta$, may be finite or infinite.

21. Some sequence spaces

The set of all sequences $\mathbf{x} = \{x_n\}$ of real or complex numbers can be turned into a metric space by taking

$$\rho(\mathbf{x}, \mathbf{y}) = \sum_1^\infty A_n \frac{|x_n - y_n|}{1+|x_n-y_n|},$$

where ΣA_n is any convergent series of positive terms. This function satisfies all the conditions for a metric; as we saw in §15, $|x-y|/\{1+|x-y|\}$ satisfies the triangle inequality since $|x-y|$ does. Since $\rho(\mathbf{x},\mathbf{y}) < \sum_1^\infty A_n$, the space is bounded.

The set of all bounded sequences $\mathbf{x} = \{x_n\}$ of real or complex numbers can be metrized by

$$\rho(\mathbf{x}, \mathbf{y}) = \sup\{|x_n - y_n| : n = 1, 2, 3, \ldots\}.$$

But this metric space is unbounded, since for every positive number k, $\rho(k\mathbf{x}, k\mathbf{y}) = k\rho(\mathbf{x}, \mathbf{y})$.

A more difficult example of a sequence space consists of all sequences $\mathbf{x} = \{x_n\}$ of real or complex numbers such that $\Sigma|x_n|^p$ is convergent for some fixed real value of $p > 1$. For this sequence space

$$\rho(\mathbf{x}, \mathbf{y}) = \left\{\sum_1^\infty |x_n - y_n|^p\right\}^{1/p}$$

is a metric.

We have first to show that $\rho(\mathbf{x}, \mathbf{y})$ is a finite real number for every pair $\mathbf{x}, \mathbf{y}$ of points of the set. By Minkowski's inequality,

$$\begin{aligned}\left\{\sum_1^m |x_n - y_n|^p\right\}^{1/p} &\leqslant \left\{\sum_1^m (|x_n| + |y_n|)^p\right\}^{1/p} \\ &\leqslant \left\{\sum_1^m |x_n|^p\right\}^{1/p} + \left\{\sum_1^m |y_n|^p\right\}^{1/p} \\ &\leqslant \left\{\sum_1^\infty |x_n|^p\right\}^{1/p} + \left\{\sum_1^\infty |y_n|^p\right\}^{1/p}.\end{aligned}$$

Hence $\sum_1^\infty |x_n - y_n|^p$ is convergent, and $\rho(\mathbf{x}, \mathbf{y})$ has a finite value for every pair of points of the set. Moreover $\rho(\mathbf{x}, \mathbf{y}) = 0$ if and only if $\mathbf{x}$ and $\mathbf{y}$ are the same point.

Since

$$\left\{\sum_1^\infty |x_n - y_n|^p\right\}^{1/p} \leqslant \left\{\sum_1^\infty |x_n|^p\right\}^{1/p} + \left\{\sum_1^\infty |y_n|^p\right\}^{1/p},$$

the sequence $\{x_n - y_n\}$ belongs to the set whenever $\{x_n\}$ and $\{y_n\}$ do. Let $\mathbf{a} = \{a_n\}$, $\mathbf{b} = \{b_n\}$, $\mathbf{c} = \{c_n\}$ be three points of the set. Putting $x_n = a_n - b_n$, $y_n = a_n - c_n$ in the last inequaltiy, we have

$$\begin{aligned}\rho(\mathbf{b}, \mathbf{c}) &= \left\{\sum_1^\infty |b_n - c_n|^p\right\}^{1/p} \\ &\leqslant \left\{\sum_1^\infty |a_n - b_n|^p\right\}^{1/p} + \left\{\sum_1^\infty |a_n - c_n|^p\right\}^{1/p} \\ &= \rho(\mathbf{a}, \mathbf{b}) + \rho(\mathbf{a}, \mathbf{c}),\end{aligned}$$

the triangle inequality.

EXERCISES

1. E is the set of all ordered pairs $\mathbf{x} = (x_1, x_2)$ of real numbers. Prove that $\rho(\mathbf{x}, \mathbf{y}) = \max\{|x_1 - y_1|, |x_2 - y_2|\}$ is a metric on E.

2. E is the set of all functions $x(t)$ continuous on the bounded closed interval $a \leqslant t \leqslant b$. Prove that

$$\rho(x, y) = \sup\{|x(t) - y(t)| : t \in [a, b]\}$$

is a metric on E.

3. E is the set of all differentiable functions $x(t)$ whose derivatives are continuous on the bounded closed interval $a \leqslant t \leqslant b$. Prove that

$$\rho(x, y) = |x(a) - y(a)| + \sup\{|x'(t) - y'(t)| : t \in [a, b]\}$$

is a metric on E.

4. $\rho(x, y)$ is a metric on a set E. Is $\rho^2(x, y)$ a metric? Is $\min\{\rho(x, y), 1\}$ a metric?

5. Is the function

$$\rho(x, y) = \left| \frac{x}{1 + \sqrt{(1 + x^2)}} - \frac{y}{1 + \sqrt{(1 + y^2)}} \right|$$

a metric on the real line? If $\rho(x, +\infty)$, $\rho(x, -\infty)$, $\rho(+\infty, -\infty)$ are defined as limits, is the function a metric on the extended real line?

6. M_1 is a metric space with metric $\rho_1(x_1, y_1)$; M_2 is a metric space with metric $\rho_2(x_2, y_2)$. Prove that

(i) $\rho_1(x_1, y_1) + \rho_2(x_2, y_2)$,

(ii) $\max\{\rho_1(x_1, y_1), \rho_2(x_2, y_2)\}$,

(iii) $\sqrt{\{\rho_1^2(x_1, y_1) + \rho_2^2(x_2, y_2)\}}$

are metrics on $M_1 \times M_2$.

7. A and B are non-empty subsets of a metric space. Prove that, if $A \subset B$, then $\delta(A) \leqslant \delta(B)$.

8. A and B are non-empty subsets of a metric space. Prove that, if $A \cap B$ is not empty, then $\delta(A \cup B) \leqslant \delta(A) + \delta(B)$.

CHAPTER 3

OPEN AND CLOSED SETS

22. Spheres

If a is a point of a metric space M with metric ρ, the set of all points x of M such that $\rho(a,x) < r$, where $r > 0$, is called a *sphere* with centre a and radius r or a *spherical neighbourhood* of the point a of radius r. Although the word 'sphere' is used in elementary geometry in the sense 'spherical surface', the present usage should cause no confusion as the set of points $\rho(a,x) = r$ in a metric space occurs rarely. Some authors call a spherical neighbourhood a *ball*, the translation of the French term *boule*; but there is no particular advantage in this, since a ball may be hollow like a ping-pong ball or solid like a golf ball. The reader is warned that both terms, sphere and ball, will be found in the literature.

In the theory of functions of a complex variable, the 'sphere' $|z-a| < r$ is often called a *disc*. Again, there is no reason for using the term 'disc' for a spherical neighbourhood in a metric space. Although no question of dimensionality arises, a word suggesting a flat object might be misleading. There is, in fact, no really satisfactory word; so we content ourselves by using 'sphere'.

The sphere with centre a and radius r will be denoted by $N(a;\, r)$, the letter being used as an abbreviation for 'neighbourhood'.

If M is the real line with metric $|x-y|$, $N(a;\, r)$ is the set of all real numbers such that $a-r < x < a+r$. But if M is the set of real numbers x such that $0 \leqslant x \leqslant 1$, $N(\frac{1}{2};\, \frac{1}{4})$ is the set $\frac{1}{4} < x < \frac{3}{4}$; but $N(0;\, \frac{1}{4})$ is the set $0 \leqslant x < \frac{1}{4}$, as also is $N(\frac{1}{16};\, \frac{3}{16})$. The last two results look rather queer; but it must be remembered that in them, M consists only of real numbers for which $0 \leqslant x \leqslant 1$.

The Euclidean plane consists of all ordered pairs

$$\mathbf{x} = (x_1, x_2), \quad \mathbf{y} = (y_1, y_2), \ldots$$

of real numbers with metric

$$\rho(\mathbf{x}, \mathbf{y}) = \sqrt{\{(x_1-y_1)^2+(x_2-y_2)^2\}},$$

and the spheres are ordinary circular discs. But if we impose on the set $R \times R$ of all ordered pairs of real numbers the metric

$$\rho(\mathbf{x}, \mathbf{y}) = \max\{|x_1-y_1|, |x_2-y_2|\},$$

the spheres are squares $|x_1-a_1| < r$, $|x_2-a_2| < r$. And if we impose the discrete metric on the same set, each sphere of radius not exceeding unity consists of a single point, every sphere of radius exceeding unity is the whole metric space.

The set of all points x of M such that $\rho(a, x) \leqslant r$, where $r > 0$ is called a *closed sphere* with centre a and radius r, and will be denoted by $K(a; r)$.

23. Open sets

If A is a subset of a metric space M, the point a of A is said to be an *interior point* of A if a is the centre of a sphere which consists only of points of A. The set of all interior points of A is called the *interior* of A and is denoted by $\operatorname{Int} A$ or $\mathring{A}$. If every point of A is an interior point of A, A is called an *open set*. For example, on the real line, the set of points x for which $h < x < k$ is an open set, and is usually denoted by (h, k). Again, the set of points of the real line for which $x > h$ is also open: it is denoted by $(h, +\infty)$.

A sphere is an open set.

If b is a point of the sphere $N(a; r)$, then $\rho(a, b) = r' < r$. If x belongs to $N(b; r-r')$, we have

$$\rho(a,x) \leqslant \rho(a,b)+\rho(b,x) < r'+(r-r') = r$$

so that $N(b; r-r') \subseteq N(a,r)$. Hence b is an interior point of $N(a; r)$, which proves the result. This includes the result for the real line, since the interval $(a-r, a+r)$ is the sphere $N(a, r)$.

The empty set is an open set.

To prove this, we have to show that $\varnothing$ contains no point which is not an interior point; this is true since $\varnothing$ contains no points.

If M is a metric space, M is open.

This is obvious, since every point is a point of M.

A subset O of a metric space M is an open set if and only if it is the union of a family of spheres.

The empty set is an open set, and it is the union of an empty family of spheres. Conversely, the union of an empty family of spheres is empty, and so is open.

If O is a non-empty open set, each point of O is the centre of a sphere contained in O, and O is the union of the family of all such spheres. Conversely, if O is the union of a non-empty family $\mathscr{S}$ of spheres, and if b is any point of O, b lies in a sphere $N(a; r)$ belonging to $\mathscr{S}$. If b is not the centre of this sphere, then it is the centre of a sphere $N(b; r')$ contained in $N(a; r)$, and hence contained in O. Thus, in either case, b is an interior point of O, which was to be proved.

The union of a family of open sets in a metric space is open.

Each open set of the family is the union of a family of spheres. Hence the union of the family of open sets is the union of a family of spheres, which proves the result.

The intersection of a finite number of open sets in a metric space is open.

The word 'finite' is essential in this enunciation. For consider the metric space consisting of the real line with metric $|x-y|$. The interval $0 < x < 1+1/n$ is open for every positive integer n. But the intersection of this infinite family is the interval $0 < x \leqslant 1$, which is not an open set.

Let O_1 and O_2 be two open sets. If O_1 or O_2 is empty, or if O_1 and O_2 are disjoint, $O_1 \cap O_2$ is the empty set which is open. So we have only to consider the case when none of $O_1, O_2, O_1 \cap O_2$ is empty.

Let $O_1 \cap O_2 = O$. If a is any point of O, it is a point of the open set O_1. Thus a is the centre of a sphere $N(a; r_1)$ contained in O_1; similarly a is the centre of a sphere $N(a; r_2)$ contained in O_2. The sphere $N(a; r)$ where r is the smaller of r_1 and r_2, is therefore contained in O_1 and in O_2, and this implies that $N(a;r)$ is contained in O. This proves that a is an interior point of O and hence that O is open. The result for the intersection of any finite number of open sets follows by induction.

The interior of any subset A of a metric space M is the largest open set contained in A.

If A has no interior points, its interior is the empty set which is open, and A contains no other open set.

If the interior of A is not empty, let a be an interior point of A, so that there is a sphere $N(a; r_1)$ contained in A. If b is any point of $N(a; r_1)$, there is a sphere $N(b; r_2)$ contained in $N(a; r_1)$ and hence in A. Therefore b is also an interior point of A. The interior of A is therefore the union of a family of spheres, and so is an open set.

If B is any open set contained in A, B is a subset of the interior of A. Hence the interior of A is the largest open subset of A.

24. Adherent points

If A is a subset of a metric space M, the point a of M is said to be an *adherent point* of A if every sphere with centre a contains a point of A. Adherent points are of two types.

An adherent point a of a subset A of M is called an *isolated point* if there is a sphere with centre a which contains no point of A other than a itself. If every point of A is an isolated point, A is called an isolated set.

An adherent point of A which is not an isolated point is called a *point of accumulation* of A. If a is a point of accumulation of A, every sphere with centre a contains a point of A distinct from a, and this implies that every sphere with centre a contains an infinite number of points of A. For the sphere $N(a; r)$ contains a point a_1 of A distinct from a. If $\rho(a, a_1) = r_1$, the sphere $N(a; r_1)$ contains a point a_2 of A distinct from a and a_1. And so on indefinitely. It should be noted that a point of accumulation of A is not necessarily a point of A.

The point a in a metric space M is an adherent point of the subset A if and only if $d(a, A) = 0$.

Since $d(a, A) = \inf\{\rho(x, a) : x \in A\}$, the equation $d(a, A) = 0$ implies that every sphere $N(a; r)$ contains a point of A; hence a is an adherent point of A. Conversely, if a is an adherent point of A, it is either an isolated point of A or a point of accumulation. In either case, $d(a, A) = 0$.

25. Closure

If A is a subset of a metric space M, the *closure* of A, denoted by $\bar{A}$, is the set of all adherent points of A. Thus if a belongs to $\bar{A}$, every sphere with centre a contains a point of A; and conversely.

A is contained in $\bar{A}$. If $A = \bar{A}$, we say that A is *closed*. For example, on the real line, the set $a \leqslant x \leqslant b$ is closed; it is denoted by $[a, b]$.

The closure of any subset of a metric space is closed.

Let A be any subset of the metric space M. Let $\bar{A}$ be the closure of A. Let $\bar{\bar{A}}$ be the closure of $\bar{A}$. We know that $\bar{A} \subseteq \bar{\bar{A}}$, so that we have to prove that $\bar{\bar{A}} \subseteq \bar{A}$.

Suppose that a belongs to $\bar{\bar{A}}$. Then for every positive value of ϵ, there exists a point b of $\bar{A}$ such that $\rho(a, b) < \epsilon$. But since b is a point of $\bar{A}$, there exists a point c of A such that

$$\rho(b, c) < \epsilon - \rho(a, b).$$

Hence
$$\rho(a, c) \leqslant \rho(a, b) + \rho(b, c) < \epsilon.$$

This means that, for every positive value of ϵ, the sphere $N(a; \epsilon)$ contains a point c of A. Hence a is a point of $\bar{A}$. Thus $\bar{\bar{A}} \subseteq \bar{A}$, which was to be proved.

If A and B are subsets of a metric space such that $A \subseteq B$, then $\bar{A} \subseteq \bar{B}$.

This is evident, since every adherent point of A is an adherent point of B.

If A and B are subsets of a metric space, then

$$\overline{A \cup B} = \bar{A} \cup \bar{B}, \quad \overline{A \cap B} \subseteq \bar{A} \cap \bar{B}.$$

An adherent point of A is an adherent point of $A \cup B$, so that $\bar{A} \subseteq \overline{A \cup B}$; similarly $\bar{B} \subseteq \overline{A \cup B}$. Hence $\bar{A} \cup \bar{B} \subseteq \overline{A \cup B}$. Thus we have to prove that $\overline{A \cup B} \subseteq \bar{A} \cup \bar{B}$.

Let c be any point of $\overline{A \cup B}$. Suppose that c does not belong to $\bar{A} \cup \bar{B}$. Since c does not belong to $\bar{A}$, there is a sphere $N(c; r_1)$ which contains no point of A; similarly there is a sphere $N(c; r_2)$ which contains no point of B. Therefore, if r is the smaller of r_1 and r_2, the sphere $N(c; r)$ contains no point of $A \cup B$. This is impossible, since c is by definition an adherent point of $A \cup B$, and so it is not the case that c does not belong to $\bar{A} \cup \bar{B}$. Therefore $\overline{A \cup B} \subseteq \bar{A} \cup \bar{B}$.

More generally, the closure of the union of any finite family of subsets of M is the union of their closures. For an infinite family, it is not necessarily the case that the closure of the union is the union of the closures of the subsets of the family. If A_α is an infinite family of subsets of M, with index set S, all we can assert in general is that

$$\bigcup_{\alpha \in S} \bar{A}_\alpha \subseteq \overline{\bigcup_{\alpha \in S} A_\alpha}.$$

To prove that $\overline{A \cap B} \subseteq \bar{A} \cap \bar{B}$, we observe that $A \cap B \subseteq A$. Hence $\overline{A \cap B} \subseteq \bar{A}$, and similarly $\overline{A \cap B} \subseteq \bar{B}$. Therefore

$$\overline{A \cap B} \subseteq \bar{A} \cap \bar{B}.$$

More generally, the closure of the intersection of any family, finite or infinite, of subsets of a metric space is contained in the intersection of their closures. It should be noted that it can happen that $\overline{A \cap B} \subset \bar{A} \cap \bar{B}$. For A and B may be disjoint, so that $\overline{A \cap B} = \bar{\varnothing} = \varnothing$, and yet $\bar{A} \cap \bar{B}$ is not empty. For example, if A is the set of all rational numbers, B the set of all irrational numbers, $A \cap B = \varnothing$; but $\bar{A} = \bar{B} = R$.

26. Interior, Exterior and Frontier Points

It is sometimes more convenient to write A^- for the closure of the subset A. Then A'^- would be the closure of the complement of A, but $A^{-\prime}$ the complement of the closure.

a is an interior point of a subset A if there exists a sphere $N(a; r)$ contained in A. The *interior* of A, denoted by $\operatorname{Int} A$ or $\mathring{A}$, is the set of all interior points of A, an open set.

A point is said to be an *exterior point* of A if it is an interior point of the complement A'. The *exterior* of A is the set of all exterior points of A, and is therefore the interior of A'. The exterior of A, denoted by $\operatorname{Ext} A$, is therefore also an open set, and

$$\operatorname{Ext} A = \operatorname{Int} A', \quad \operatorname{Ext} A' = \operatorname{Int} A.$$

The point a is an exterior point of A if and only if there is a sphere $N(a; r)$ which contains no point of A, that is, if and only if a is not an adherent point of A. Therefore the exterior of A is the complement of the closure of A,

$$\operatorname{Ext} A = A^{-\prime}.$$

It follows that $$\operatorname{Int} A = A^{\prime-\prime}.$$

The exterior and interior of A do not necessarily fill up the whole space M. For

$$(\operatorname{Ext} A \cup \operatorname{Int} A)' = (\operatorname{Ext} A)' \cap (\operatorname{Int} A)'$$
$$= A^{-\prime\prime} \cap A^{\prime-\prime\prime} = A^{-} \cap A^{\prime-}.$$

The set $\bar{A} \cap \overline{A'}$ is not necessarily empty; it is called the *frontier* of A and is denoted by Fr A (or sometimes by $\dot{A}$ or ∂A); A and A' have the same frontier. The frontier of A is the complement of the open set $\operatorname{Int} A \cup \operatorname{Ext} A$, and this, as we shall see later, implies that the frontier of A is a closed set.

The *boundary* of a set A, denoted by Bd A, is the part of the frontier of A which belongs to A; hence

$$\operatorname{Bd} A = A \cap A^{\prime-}, \quad \operatorname{Bd} A' = A' \cap A^{-}.$$

Evidently, if A is closed, its frontier is its boundary.

The boundary of a set contains no non-empty open set.

For if Bd A contained a non-empty open set O, O would be contained in A. Since Int A is the largest open subset of A, O would be contained in Int A. This is impossible since

$$O \subseteq \operatorname{Bd} A \subseteq \operatorname{Fr} A \subseteq (\operatorname{Int} A)'.$$

Hence there is no non-empty open set contained in Bd A. The frontier of a set does not necessarily have this property; for example, on the real line, the frontier of the set of all rational numbers is the whole line.

27. Dense subsets

A subset A of a metric space M is said to be *everywhere dense* in M if M is the closure of A. If the exterior of A is everywhere dense in M, A is said to be *nowhere dense* in M.

The subset A of a metric space M is nowhere dense in M if and only if the closure of A has no interior points.

If A is nowhere dense, the closure of Ext A is M, that is

$$A^{-\prime-} = M.$$

Taking complements, $$A^{-\prime-\prime} = \varnothing$$

and so $$\operatorname{Int}(\bar{A}) = \varnothing.$$

And conversely.

A metric space M is said to be *separable* if there is a countable subset everywhere dense in M. For example, the real line is separable; the closure of the countable set of all rational numbers is the real line.

A subset A of a metric space M is said to be dense with respect to another subset B if $B \subseteq \bar{A}$. Evidently every subset is dense with respect to itself. This is not the same as saying that every subset is dense-in-itself. A subset A of a metric space M is *dense-in-itself* when every point of A is a point of accumulation of A. A subset which is both closed and dense-in-itself is said to be *perfect*.

28. Closed sets

Just as we used the letter O to denote an open set (*un ensemble ouvert*), so we shall use the letter F to denote a closed set (*un ensemble fermé*). There are other conventions; this seems a convenient one.

It follows from the definition that, *in a metric space M, every finite set, the empty set and the whole space are closed sets.* It will be noticed that M and $\varnothing$ are both open and closed. Sometimes there are other subsets of M with this property. For example, consider the metric space M consisting of all complex numbers such that $|z-1| < 1$ or $|z+1| < 1$, with metric $\rho(z_1, z_2) = |z_1 - z_2|$. Then $|z-1| < 1$ is an open subset of M. By a theorem we shall prove shortly, the complement of $|z-1| < 1$ with respect to M is closed. This complement is $|z+1| < 1$ which is also open. Thus $|z+1| < 1$, and similarly $|z-1| < 1$, are subsets of M which are both open and closed. Naturally $|z+1| < 1$ is not a closed set in the whole complex plane: that is not the point—it is a closed subset of the metric space M.

The complement of a closed set is open; the complement of an open set is closed.

This theorem states that a set is closed if and only if its complement is open. For example, on the real line, the set $x < 0$ is open, and hence its complement $x \geqslant 0$ is closed. But if the space consists only of the real numbers such that $-1 \leqslant x \leqslant 1$, the set $0 \leqslant x \leqslant 1$ is closed, its complement $-1 \leqslant x < 0$ is open.

The theorem is true for the whole space M and for the empty set $\varnothing$, since $M' = \varnothing$, $\varnothing' = M$ and M and $\varnothing$ are each both open and closed. Accordingly we disregard these cases.

Let F be a closed set. Let a be any point of its complement F'. There must be a sphere $N(a;\, r)$ which contains no point of F; for if every such sphere contained a point of F, a would be an adherent point of F, and therefore a point of F since F is closed. The point a is thus an interior point of F'. Therefore all the points of F' are interior points, which means that F' is open.

Let O be an open set. Let a be any adherent point of its complement O'. The point a cannot belong to O; for if it did, there would be a sphere $N(a;\, r)$ contained in O, which is impossible. Hence every adherent point of O' belongs to O', and so O' is closed.

A closed sphere is a closed set.

The closed sphere $K(a;r)$ with centre a and radius $r > 0$ was defined to be the set of all points x such that $\rho(x, a) \leqslant r$. If

$$y \in K'(a;\, r),$$

then $\rho(y, a) > r$. If $\rho(y, a) = r + 2\epsilon$, the sphere $N(y;\, \epsilon)$ belongs to $K'(a;\, r)$ since every point of $N(y;\, \epsilon)$ is at a distance greater than $r + \epsilon$ from a. The set $K'(a;\, r)$ is therefore the union of a family of spheres, and so is open; hence $K(a;\, r)$ is closed.

Since $N(a;\, r) \subseteq K(a;\, r)$, we have

$$\overline{N(a;\, r)} \subseteq \overline{K(a;\, r)} = K(a;\, r).$$

Thus the closure of $N(a;\, r)$ is contained in $K(a;\, r)$, but may be different from $K(a;\, r)$. For example, if the metric space consists of all real numbers x such that $x \leqslant 0$ and the positive integers, with the metric $|x-y|$, we have

$$\overline{N(0;\,1)} = \{x \in R: -1 \leqslant x \leqslant 0\},$$

$$K(0;\,1) = \{x \in R: -1 \leqslant x \leqslant 0\} \cup \{1\},$$

so that the closure of $N(0;\,1)$ is a proper subset of $K(0;\,1)$.

The intersection of a family of closed sets in a metric space is closed.

Let F_α be a family of closed sets with index set S. The complement of their intersection,

$$(\bigcap_{\alpha \in S} F_\alpha)' = \bigcup_{\alpha \in S} F'_\alpha,$$

is open, being the union of a family of open sets. Hence $\bigcap F_\alpha$ is closed. Similarly we can show that

The union of a finite number of closed sets in a metric space is closed.

The word 'finite' is essential here. For example, if $\{F_n\}$ is the sequence of closed subsets $1/n \leqslant x \leqslant 1$ of the real line, their union

$$\bigcup_{n=1}^{\infty} F_n$$

is the set $0 < x \leqslant 1$ which is not a closed subset of the real line.

The closure of a subset A of a metric space M is the intersection of all closed sets containing A.

There are closed sets containing A; for instance $\bar{A}$ and M. Let F be the intersection of all closed sets containing A; then F is closed. But since $A \subseteq F$, $\bar{A} \subseteq \bar{F} = F$. Hence every closed set which contains A contains $\bar{A}$. But $\bar{A}$ is a closed set containing A, while F, being the intersection of all closed sets containing A, is contained in each of them. In particular $F \subseteq \bar{A}$. Therefore $\bar{A} = F$. The closure of A is thus the least closed set containing A.

29. Subspaces of a metric space

If M_1 is a subspace of M, a set which is open in M_1 is not necessarily open in M; a set which is closed in M_1 is not necessarily closed in M. For example, if M is the real line R with the usual metric and M_1 the subspace $0 \leqslant x \leqslant 1$, the subset $0 \leqslant x < \frac{1}{2}$ is open in M_1 but not in M.

In order that $B \subseteq M_1$ be open in the subspace M_1 of a metric space M, it is necessary and sufficient that there exists a set O, open in M, such that $B = O \cap M_1$.

In other words, the open sets in M_1 are the intersections of M_1 with the open sets of M.

If a is a point of M_1, denote by $N_1(a; r)$ the set of points of M_1 such that $\rho(a, x) < r$. The sets $N_1(a; r)$ are the spheres of M_1 since the metric of M_1 is the induced metric. Evidently

$$N_1(a; r) = M_1 \cap N(a, r).$$

Let B be a subset of M_1, open relative to M_1. Then if x is any point of B, there is a sphere $N_1(x; r(x))$ contained in B, where the notation indicates that the radius $r(x)$ depends on x. We then have

$$B = \bigcup_{x \in B} N_1(x; r(x)) = M_1 \cap \bigcup_{x \in B} N(x; r(x)) = M_1 \cap O,$$

where

$$O = \bigcup_{x \in B} N(x; r(x)),$$

being the union of a family of spheres of M, is open with respect to M. This proves the condition is necessary.

Conversely, let $B = O \cap M_1$ where O is open relative to M. If x is any point of B, x is a point of O, and so there exists a sphere $N(x; r(x))$ contained in O. Hence

$$N_1(x; r(x)) = M_1 \cap N(x; r(x)) \subseteq M_1 \cap O = B$$

so that x is an interior point of the subset B of the subspace M_1. As x was any point of B, B is open relative to M_1, and so the condition is sufficient.

In order that every subset B of the subspace M_1, which is open in M_1, be open in M, it is necessary and sufficient that M_1 be open in M.

M_1 is open in M_1, so that the condition is necessary.

If B is open in M_1, $B = O \cap M_1$ where O is open in M. If M_1 is open in M, then B, being the intersection of two open sets, is open in M. Thus the condition is sufficient.

In order that $B \subseteq M_1$ be closed in the subspace M_1 of the metric space M, it is necessary and sufficient that there exist a set F, closed in M, such that $B = F \cap M_1$.

B is closed in M_1 if and only if $B' \cap M_1$ is open in M_1. Hence B is closed in M_1 if and only if there exists a set O, open in M, such that

$$B' \cap M_1 = O \cap M_1.$$

If this condition is satisfied, we have $B \cup M_1' = O' \cup M_1'$ by taking complements. Hence

$$B = B \cap M_1 = (B \cup M_1') \cap M_1 = (O' \cup M_1') \cap M_1 = O' \cap M_1,$$

so that B is the intersection of M_1 and the closed set O'.

Conversely, let $B = F \cap M_1$, where F is a closed subset of M. Then $B' = F' \cup M_1'$, and so

$$B' \cap M_1 = (F' \cup M_1') \cap M_1 = F' \cap M_1,$$

where F' is open in M. Hence the condition is satisfied.

In order that every subset B of the subspace M_1, which is closed in M_1, be closed in M, it is necessary and sufficient that M_1 be closed in M.

M_1 is closed in M_1, so that the condition is necessary. If B is closed in M_1, $B = F \cap M_1$ where F is closed in M. If M_1 is closed in M, B, being the intersection of two closed sets, is closed in M. Thus the condition is sufficient.

The closure of a subset B of the subspace M_1 of a metric space M with respect to M_1 is $\bar{B} \cap M_1$.

If a is an adherent point of B with respect to M_1, every sphere in M_1, $N_1(a; r)$ say, contains a point of B. But

$$N_1(a; r) = N(a; r) \cap M_1 \subseteq N(a, r).$$

Hence a is an adherent point of B with respect to M. Thus $a \in \bar{B} \cap M_1$. The closure of B with respect to M_1 is contained in $\bar{B} \cap M_1$.

Conversely, if $a \in \bar{B} \cap M_1$, it is evidently an adherent point of B with respect to M_1. Hence $\bar{B} \cap M_1$ is contained in the closure of B with respect to M_1. This completes the proof.

EXERCISES

1. $N(a; r)$ is a sphere in the metric space M. A is a subset of M which intersects $N(a; r)$ and has diameter less than r. Prove that $A \subseteq N(a; 2r)$.

2. Prove that the diameter of a subset of a metric space is equal to the diameter of its closure.

3. M is a metric space consisting of a set E with metric ρ. M_1 is the metric space consisting of the same set E with metric ρ_1. The metrics ρ and ρ_1 are said to be equivalent if the open sets of M are the open sets of M_1 and conversely. Prove that the metrics ρ and $\rho/(1+\rho)$ are equivalent.

4. The Euclidean plane is the set of all ordered pairs $\mathbf{x} = (x_1, x_2)$ of real numbers, with metric $\rho(\mathbf{x}, \mathbf{y}) = \sqrt{\{(x_1-y_1)^2+(x_2-y_2)^2\}}$. The function

$$\rho_1(\mathbf{x}, \mathbf{y}) = \max\{|x_1-y_1|, |x_2-y_2|\}$$

is also a metric on the same set. Prove that ρ and ρ_1 are equivalent.

5. A is a set of points in a metric space M; B is the set of all points of accumulation of A. Prove that B is closed.

6. A, B, C are subsets of a metric space M; A is dense with respect to B, B is dense with respect to C. Prove that A is dense with respect to C.

7. Prove that the following results are equivalent:
   (i) A is everywhere dense in the metric space M.
   (ii) The only closed set which contains A is M.
   (iii) The only open set disjoint from A is the empty set.
   (iv) A intersects every non-empty open set.
   (v) A intersects every sphere.

8. Show that a subset A of a metric space M is nowhere dense if and only if every sphere in M contains a sphere free from points of A.

9. Show that any subspace of a separable metric space is separable.

10. A and B are disjoint closed subsets of a metric space M. Show that

$$U = \{x \in M: d(x, A) < d(x, B)\}, \quad V = \{x \in M: d(x, B) < d(x, A)\}$$

are disjoint open sets containing A and B respectively.

11. O is an open set in a metric space M; A is any subset of M. Prove that

$$O \cap \bar{A} \subseteq \overline{O \cap A}, \quad \overline{O \cap \bar{A}} = \overline{O \cap A}.$$

Deduce that, if $O \cap A$ is empty, so also is $O \cap \bar{A}$.

12. If A and B are two subsets of a metric space M,

$$\mathrm{Fr}(A \cup B) \subseteq \mathrm{Fr}\,A \cup \mathrm{Fr}\,B.$$

Show also that, if $\bar{A}$ and $\bar{B}$ are disjoint, $\mathrm{Fr}(A \cup B) = \mathrm{Fr}\,A \cup \mathrm{Fr}\,B$.

13. Prove that the complement of the interior of a subset A of a metric space M is the closure of the complement of A.

Show that
   (i) $\mathrm{Fr}\,\bar{A} \subseteq \mathrm{Fr}\,A$,
   (ii) $\mathrm{Fr}\,\mathrm{Int}\,A \subseteq \mathrm{Fr}\,A$,
   (iii) $\bar{A} = \mathrm{Int}\,A \cup \mathrm{Fr}\,A$.

14. Prove that

$$\mathrm{Int}(A \cap B) = \mathrm{Int}\,A \cap \mathrm{Int}\,B,$$

$$\mathrm{Ext}(A \cup B) = \mathrm{Ext}\,A \cap \mathrm{Ext}\,B.$$

15. Prove that the frontier of a subset A of a metric space M is empty if and only if A is both open and closed.

16. Prove that an open set contains none of its frontier points.

17. If O_1 and O_2 are two open sets in a metric space M, prove that

$$(O_1 \cap \operatorname{Fr} O_2) \cup (O_2 \cap \operatorname{Fr} O_1) \subseteq \operatorname{Fr}(O_1 \cap O_2)$$
$$\subseteq (O_1 \cap \operatorname{Fr} O_2) \cup (O_2 \cap \operatorname{Fr} O_1) \cup (\operatorname{Fr} O_1 \cap \operatorname{Fr} O_2).$$

18. If A is any subset of a metric space M, the sets $\alpha(A)$, $\beta(A)$ are defined by

$$\alpha(A) = \operatorname{Int} \bar{A}, \quad \beta(A) = \overline{\operatorname{Int} A}.$$

Prove that $\alpha\alpha(A) \subseteq \alpha(A), \beta\beta(A) \supseteq \beta(A)$.

If O is an open set, F a closed set, prove that $O \subseteq \alpha(O)$, $F \supseteq \beta(F)$. Deduce that, for any set A, $\alpha\alpha(A) = \alpha(A), \beta\beta(A) = \beta(A)$.

19. A and B are non-empty subsets of a bounded metric space M with metric ρ; $d(x, A)$ and $d(x, B)$ are the distances of a point x of M from A and B respectively. If x is a given point of M, show that, for every positive number ϵ, there exists a point b of B and a point a of A such that

$$\rho(b, x) \leqslant d(x, B) + \epsilon, \quad \rho(a, b) \leqslant d(b, A) + \epsilon.$$

Deduce that, if $\rho_{AB} = \sup\{d(x, A) : x \in B\}$, then

$$d(x, A) \leqslant \rho_{AB} + d(x, B) + 2\epsilon.$$

Hence show that, if C is a third non-empty subset of M, then

$$\rho_{AC} \leqslant \rho_{AB} + \rho_{BC}.$$

$\mathscr{F}$ is the family of all closed non-empty subsets of M. Show that $\max(\rho_{AB}, \rho_{BA})$ is a metric on $\mathscr{F}$.

20. M is any set, and ρ is a real valued non-negative function such that, for all members x, y, z of M, $\rho(x, y) = 0$ if and only if $x = y$ and

$$\rho(x, y) \leqslant \max(\rho(x, z), \rho(y, z)).$$

Show that ρ is a metric on M.

Show that, in this metric space, every sphere $N(a; r)$ is both open and closed, and that every closed sphere $K(a; r)$ is both open and closed. Prove also that, if $K(a; r)$ and $K(b; s)$ have a point in common, one of these closed spheres is contained in the other.

21. The metric space M_1 is the set E_1 with metric ρ_1; the metric space M_2 is the set E_2 with metric ρ_2. The metric space $M_1 \times M_2$ is defined to be the set $E_1 \times E_2$ with metric

$$\rho(\mathbf{x}, \mathbf{y}) = \max\{\rho_1(x_1, y_1), \rho_2(x_2, y_2)\}$$

where $\mathbf{x} = (x_1, x_2)$, $\mathbf{y} = (y_1, y_2)$ are any two points of $E_1 \times E_2$. If $\mathbf{a} = (a_1, a_2)$ is any point of $M_1 \times M_2$ and if $r > 0$, prove that

$$N(\mathbf{a}; r) = N_1(a_1; r) \times N_2(a_2; r),$$

$$K(\mathbf{a}; r) = K_1(a_1; r) \times K_2(a_2; r).$$

Deduce that, if A_1 is open in M_1, A_2 open in M_2, then $A_1 \times A_2$ is open in $M_1 \times M_2$.

Show also that, if A_1 is a subset of M_1, A_2 a subset of M_2, the closure of $A_1 \times A_2$ is $\bar{A}_1 \times \bar{A}_2$, and that a necessary and sufficient condition for $A_1 \times A_2$ to be closed in $M_1 \times M_2$ is that A_1 be closed in M_1, and A_2 be closed in M_2.

CHAPTER 4

COMPLETE METRIC SPACES

30. Convergent sequences

The sequence of points $\{a_n\}$ in a metric space M is said to converge to a point a of M if the distance $\rho(a_n, a)$ tends to zero as $n \to \infty$, that is, if for every positive value of ϵ there exists an integer n_0, depending on ϵ, such that

$$0 \leqslant \rho(a_n, a) < \epsilon,$$

whenever $n \geqslant n_0$. The point a is called the limit of the sequence. We say the limit, because no sequence can converge to two limits; for if $\{a_n\}$ converged to a and to b, we should have

$$0 \leqslant \rho(a, b) \leqslant \rho(a_n, a) + \rho(a_n, b) \to 0$$

as $n \to \infty$, and so $\rho(a, b) = 0$ which is impossible since a and b are distinct.

In particular, if $a_n = a$ for all but a finite number of values of n, the sequence $\{a_n\}$ converges to a.

If $\{k_n\}$ is a strictly increasing sequence of positive integers, so that $k_n \geqslant n$ and $k_n \to \infty$ as $n \to \infty$, the sequence $\{a_{k_n}\}$ is called a subsequence of $\{a_n\}$; and if $a_n \to a$, $a_{k_n} \to a$.

We recall that a sequence is not a set. The elements in a sequence are ordered and are not necessarily distinct; the elements in a set are not ordered and are distinct. For example, we could define a sequence by $a_{2n} = b$, $a_{2n+1} = c$, where $b \neq c$; the values taken by a_n form a set with only two members.

If a is a point of accumulation of a subset A of a metric space M, there is a sequence $\{a_n\}$ of points of A, all distinct from a, which converges to a.

Every sphere with centre a contains a point of A distinct from a (which may or may not belong to A). Start with any point a_1 of A, and let $r_1 = \min(1, \rho(a, a_1))$. Then the sphere $N(a; r_1)$ contains a point a_2 of A, distinct from a. Let $r_2 = \min(\frac{1}{2}, \rho(a, a_2))$; the sphere $N(a; r_2)$ contains a point a_3 of A, distinct from a.

Let $r_3 = \min(\frac{1}{3}, \rho(a, a_3))$; the sphere $N(a; r_3)$ contains a point a_4 of A, distinct from a. And so on indefinitely. In this way, we construct a sequence of distinct points $\{a_n\}$, all distinct from a, such that, if $r_n = \min(1/n, \rho(a, a_n))$, the sphere $N(a; r_n)$ contains the point a_{n+1} of A. Thus $\rho(a, a_n) < r_{n-1} \leqslant 1/(n-1)$, and so $\rho(a, a_n)$ tends to zero as $n \to \infty$. Hence the sequence $\{a_n\}$ of points of A converges to the point of accumulation a.

If a and b are two points of a metric space and if $\{b_n\}$ is a sequence which converges to b, $\rho(a, b_n)$ converges to $\rho(a, b)$.

By the triangle inequality,

$$|\rho(x, y) - \rho(x, z)| \leqslant \rho(y, z).$$

If we put $x = a$, $y = b$, $z = b_n$ we have

$$|\rho(a, b) - \rho(a, b_n)| \leqslant \rho(b, b_n).$$

Since $\rho(b, b_n)$ converges to zero, $\rho(a, b_n)$ converges to $\rho(a, b)$.

If a and b are two points of a metric space, and if $\{a_n\}$ and $\{b_n\}$ are sequences which converge to a and b respectively, $\rho(a_n, b_n)$ converges to $\rho(a, b)$.

For

$$\begin{aligned} |\rho(a, b) - \rho(a_n, b_n)| &\leqslant |\rho(a, b) - \rho(a_n, b)| + |\rho(a_n, b) - \rho(a_n, b_n)| \\ &\leqslant \rho(a, a_n) + \rho(b, b_n) \end{aligned}$$

and so $\rho(a_n, b_n)$ converges to $\rho(a, b)$.

31. Cauchy's principle of convergence

One of the fundamental theorems of analysis is Cauchy's Principle of Convergence, which enables one to prove the convergence of a sequence of real numbers without knowing beforehand what the limit should be. This principle which was proved in §11 runs as follows:

The necessary and sufficient condition for the convergence of a sequence $\{a_n\}$ of real numbers is that, for every positive value of ϵ, there exists an integer $n_1(\epsilon)$ such that $|a_m - a_n| < \epsilon$ whenever $m > n \geqslant n_1$.

A sequence of real numbers $\{a_n\}$ which satisfies this condition is called a *Cauchy sequence.*

The definition of a Cauchy sequence can be extended to complex sequences, to sequences of real functions, and, indeed, to a sequence of points in any metric space. The sequence of points $\{a_n\}$ in a space with metric ρ is called a Cauchy sequence if, for every positive value of ϵ, there exists an integer $n_1(\epsilon)$ such that $\rho(a_m, a_n) < \epsilon$ whenever $m > n \geqslant n_1$.

For a sequence of points in a metric space M to be convergent, it is necessary that the sequence be a Cauchy sequence.

If the sequence $\{a_n\}$ of points of M converges to the point a of M, then, for every positive value of ϵ, there exists an integer $n_0(\epsilon)$ such that $\rho(a, a_n) < \epsilon$ whenever $n \geqslant n_0$. Hence if $m > n \geqslant n_0$,

$$0 \leqslant \rho(a_m, a_n) \leqslant \rho(a_m, a) + \rho(a, a_n) < 2\epsilon.$$

The sequence is thus a Cauchy sequence; the occurrence of 2ϵ instead of ϵ is irrelevant.

In a general metric space, a Cauchy sequence is not necessarily convergent.

For example, take the metric space consisting of all rational numbers with metric $|x-y|$. The sequence of partial sums of the series

$$\frac{1}{1!}+\frac{1}{2!}+\frac{1}{3!}+\dots$$

is a Cauchy sequence of rational numbers, but it does not converge to a rational number. Or again, take the metric space which consists of all real numbers x such that $0 < x < 1$; the sequence $\{1/n\}$ is a Cauchy sequence, but it does not converge to a point of the space.

A metric space M is said to be *complete* if every Cauchy sequence of points of M converges to a point of M.

32. Convergent Cauchy sequences

A bounded sequence of real numbers either converges or oscillates finitely; the Cauchy sequence condition ensures that it does not oscillate, and therefore that it converges. The situation is different in a general metric space. A bounded sequence may either converge or oscillate finitely or not converge at all. The Cauchy sequence condition ensures that it does not oscillate;

but, if the space is not complete, there may be no point in the space to which the sequence converges.

If a Cauchy sequence of points in a metric space contains a convergent subsequence, the sequence is convergent.

Suppose $\{a_n\}$ is a Cauchy sequence in a space M with metric ρ. Then, for every positive value of ϵ, there exists an integer $n_1(\epsilon)$ such that $\rho(a_m, a_n) < \epsilon$ whenever $m > n \geqslant n_1$.

If the sequence contains a subsequence $\{a_{k_n}\}$ which converges to the point a of M, we have

$$\rho(a_{k_m}, a_n) < \epsilon$$

whenever $m > n \geqslant n_1$, since $\{k_n\}$ is a strictly increasing sequence of positive integers. Making $m \to \infty$, we have

$$\rho(a, a_n) \leqslant \epsilon,$$

whenever $n \geqslant n_1(\epsilon)$; hence the sequence $\{a_n\}$ converges to a.

33. Cantor's intersection theorem

A necessary and sufficient condition for a metric space M to be complete is that every nest $\{K_n\}$ of closed spheres, such that the diameter of K_n tends to zero as $n \to \infty$, has a non-empty intersection.

A closed sphere $K(a; r)$ is the set of points x such that

$$\rho(x, a) \leqslant r.$$

A nest of closed spheres is a sequence of closed spheres, each of which is a proper subset of its predecessor.

Let the sphere K_n have centre x_n and radius r_n. Since

$$K_m \subset K_n \subseteq K_{n_0},$$

whenever $m > n \geqslant n_0$, the centres x_n and x_m belong to K_{n_0} and so $\rho(x_m, x_n) \leqslant 2r_{n_0}$. But since $r_n \to 0$ as $n \to \infty$, for every positive value of ϵ we can choose the integer n_0 so that $r_{n_0} < \frac{1}{2}\epsilon$. Hence $\rho(x_m, x_n) < \epsilon$ whenever $m > n \geqslant n_0$, and so the sequence $\{x_n\}$ of centres of the nest $\{K_n\}$ is a Cauchy sequence.

If M is complete, the sequence $\{x_n\}$ converges to a point x of M. But each closed sphere contains all but a finite number of points of the sequence $\{x_n\}$ and hence contains x. Therefore

$\bigcap_{n=1}^{\infty} K_n$ contains x and so is not empty; it consists of the single point x. Thus the condition is necessary.

Next suppose that the condition is satisfied and that M is not complete. Then there exists a Cauchy sequence $\{x_n\}$ which does not converge to a point of M. By the Cauchy condition, we can find a strictly increasing sequence $\{k_n\}$ of positive integers such that

$$\rho(x_m, x_{k_n}) < \frac{1}{2^{n+1}}$$

whenever $m > k_n$.

Let K_n be the closed sphere $K(x_{k_n}; 1/2^n)$. If y belongs to K_{n+1},

$$\rho(y, x_{k_{n+1}}) \leqslant \frac{1}{2^{n+1}} \quad \text{and} \quad \rho(x_{k_{n+1}}, x_{k_n}) < \frac{1}{2^{n+1}}.$$

Hence $\rho(y, x_{k_n}) < 1/2^n$ so that y is an interior point of K_n. Therefore $K_{n+1} \subset K_n$ for every positive integer n.

By hypothesis the nest $\{K_n\}$ has a non-empty intersection. There is, therefore, a point z which belongs to all the closed spheres of the nest. Since the radius of K_n tends to zero as $n \to \infty$, z is the limit of the subsequence $\{x_{k_n}\}$ of $\{x_n\}$. Hence the Cauchy sequence converges to z, contrary to the hypothesis that $\{x_n\}$ was a Cauchy sequence which did not converge to a point of M. The assumption that M is not complete is thus untenable, and so the condition is sufficient.

By a similar argument, we can prove Cantor's intersection theorem, that *if $\{F_n\}$ is a sequence of non-empty closed sets in a complete metric space M such that $F_1 \supset F_2 \supset F_3 \supset \ldots$, and if the diameter of F_n tends to zero as n tends to infinity, the intersection*

$$\bigcap_{n=1}^{\infty} F_n$$

is non-empty and consists of a single point.

34. Baire's category theorem

A subset A of a metric space M, possibly the whole space, is said to be of the first category if it is the union of a countable family of nowhere-dense sets; otherwise it is said to be of the second category. Baire's category theorem states that *a complete metric space is of the second category.*

Suppose the result untrue, so that there exists a complete metric space M which is the union of a countable family of nowhere-dense sets. We can arrange this family as a sequence $\{A_n\}$.

If A is a set nowhere-dense in M, Ext A is everywhere dense in M—the closure of Ext A is M. Hence any point x of M is an adherent point of Ext A; every sphere $N(x; r)$ contains a point y of Ext A. But since Ext A is an open set, there exists a sphere $N(y; r')$ contained in $N(x; r)$ which consists only of points of Ext A, and therefore contains no points of A. It follows that, if $0 < r'' < r'$, the closed sphere $K(y; r'')$ is contained in $N(x; r)$ but contains no points of A.

Now consider the sequence $\{A_n\}$ of nowhere-dense sets. In any sphere, we can find a closed sphere K_1, of radius $r_1 < 1$, which does not intersect A_1. In the corresponding open sphere N_1, we can find a closed sphere K_2, of radius $r_2 < \frac{1}{2}$, which does not intersect A_2. And so on. In this way we construct a nest $\{K_n\}$ of closed spheres with the following properties:

(i) for each integer n, K_n does not intersect $A_1, A_2, \ldots, A_n$;

(ii) the radius of K_n tends to zero as $n \to \infty$.

Since M is complete,

$$\bigcap_{n=1}^{\infty} K_n$$

is not empty, but consists of a single point x which does not belong to any of the nowhere-dense sets $\{A_n\}$. This is impossible since M is the union of this family. Hence the complete metric space M is not of the first category.

35. The completion of a metric space

The rational numbers do not form a complete metric space. But the rational real numbers are a subset of the metric space R of real numbers, and R is complete. The field of rational numbers is logically distinct from the field of rational real numbers, but as the two fields are isomorphic, we do not in practice need to distinguish between them. Thus we may say that it is possible to embed the field Q of rational numbers in the field R of real numbers in such a way that Q is dense in R, the embedding being isometric. We now show that it is possible to perform the same sort of completion in any metric space.

We start by considering the set of all Cauchy sequences of points of a metric space M with metric ρ, where M is not complete. On this set, we set up the relation $\{x_n\} \sim \{y_n\}$ between two Cauchy sequences as meaning that $\rho(x_n, y_n)$ tends to zero as $n \to \infty$. For example, $\{x_n\} \sim \{x_{n+p}\}$ for every fixed integer p; or again, if $\{x_n\}$ converges to x, and if $y_n = x$ for all x, $\{x_n\} \sim \{y_n\}$. This relation is an equivalence relation. The equivalence classes form a set which we denote by M^*. By an appropriate choice of the definition of distance, we can make M^* into a metric space, which turns out to be complete.

The equivalence classes which are the elements of M^* are of two types. An element of the first type is an equivalence class of convergent Cauchy sequences; if $\{x_n\}$ is a Cauchy sequence which converges to a point x of M, we say that the equivalence class which contains $\{x_n\}$ (and, of course, also contains the constant sequence all of whose members are equal to x) corresponds to the point x of M. An element of the second type is an equivalence class which does not correspond to any element of M; it consists of equivalent Cauchy sequences which do not converge to any point of M.

It is convenient to denote elements of M^* by bold letters, such as **a**, **b** or **x**. If $\{a_n\}$ and $\{b_n\}$ are Cauchy sequences contained in the equivalence classes **a** and **b** respectively, the triangle inequality gives

$$\begin{aligned}|\rho(a_m, b_m) - \rho(a_n, b_n)| &\leqslant |\rho(a_m, b_m) - \rho(a_n, b_m)| + |\rho(a_n, b_m) - \rho(a_n, b_n)| \\ &\leqslant \rho(a_m, a_n) + \rho(b_m, b_n).\end{aligned}$$

But since $\{a_n\}$ and $\{b_n\}$ are Cauchy sequences, there exists, for every positive value of ϵ, a positive integer $n_1(\epsilon)$ such that $\rho(a_m, a_n) < \epsilon$, $\rho(b_m, b_n) < \epsilon$ whenever $m > n \geqslant n_1$. Hence

$$|\rho(a_m, b_m) - \rho(a_n, b_n)| < 2\epsilon,$$

whenever $m > n \geqslant n_1$, so that $\{\rho(a_n, b_n)\}$ is a Cauchy sequence of real numbers which is therefore convergent. Moreover the limit to which $\rho(a_n, b_n)$ tends does not depend on the particular Cauchy sequences chosen from $\mathbf{a}, \mathbf{b}$. For if $\{a_n'\}$, $\{b_n'\}$ are two other Cauchy sequences belonging to $\mathbf{a}$ and $\mathbf{b}$ respectively,

$$\begin{aligned} |\rho(a_n', b_n') - \rho(a_n, b_n)| & \\ & \leqslant |\rho(a_n', b_n') - \rho(a_n, b_n')| + |\rho(a_n, b_n') - \rho(a_n, b_n)| \\ & \leqslant \rho(a_n', a_n) + \rho(b_n', b_n) \\ & \to 0 \end{aligned}$$

as $n \to \infty$, and so $\rho(a_n', b_n')$ tends to the same limit as $\rho(a_n, b_n)$. This limit, which depends only on $\mathbf{a}$ and $\mathbf{b}$, we shall denote by $\rho^*(\mathbf{a}, \mathbf{b})$. We note that, if $\mathbf{a}$ and $\mathbf{b}$ correspond to elements a and b of M, we can use the representations of $\mathbf{a}$ and $\mathbf{b}$ by constant sequences and take $a_n' = a$, $b_n' = b$; it follows that

$$\rho^*(\mathbf{a}, \mathbf{b}) = \lim_{n\to\infty} \rho(a_n', b_n') = \rho(a, b).$$

This function $\rho^*(\mathbf{a}, \mathbf{b})$ satisfies all the conditions for a metric on M^*. Evidently if $\mathbf{a} = \mathbf{b}$, $\rho^*(\mathbf{a}, \mathbf{b}) = 0$. Conversely, if

$$\rho^*(\mathbf{a}, \mathbf{b}) = 0,$$

$$\lim_{n\to\infty} \rho(a_n, b_n) = 0$$

for any Cauchy sequences $\{a_n\}$ and $\{b_n\}$ which belong to $\mathbf{a}$ and $\mathbf{b}$ respectively; hence $\{a_n\} \sim \{b_n\}$ and so $\mathbf{a} = \mathbf{b}$. Next,

$$\rho^*(\mathbf{a}, \mathbf{b}) = \rho^*(\mathbf{b}, \mathbf{a})$$

since $\rho(a_n, b_n) = \rho(b_n, a_n)$. Lastly if $\{c_n\}$ is a Cauchy sequence belonging to the element $\mathbf{c}$ of M^*,

$$\rho(b_n, c_n) \leqslant \rho(a_n, b_n) + \rho(a_n, c_n)$$

and so
$$\rho^*(\mathbf{b}, \mathbf{c}) \leqslant \rho^*(\mathbf{a}, \mathbf{b}) + \rho^*(\mathbf{a}, \mathbf{c})$$

the triangle inequality of M^*.

Let us denote by A^* the subset of elements of M^* which correspond to points of M. The restriction of ρ^* to A^* is a metric on A^*, which is a subspace of M^*. If the points $\mathbf{a}$, $\mathbf{b}$ of A^* correspond to the points a and b of M, we have proved that

$$\rho^*(\mathbf{a}, \mathbf{b}) = \rho(a, b),$$

so that A^* is isometric to M.

The subspace A^* is dense in M^*. To prove this we must show that every point of M^*, which is not a point of A^*, is a point of accumulation of A^*. Let $\mathbf{a}$ be a point of M^* which is not a point of A^*, and let $\{a_n\}$ be a Cauchy sequence of points of M belonging to the equivalence class $\mathbf{a}$. We define a sequence $\{\mathbf{a}_n\}$ of points of A^*, where the equivalence class $\mathbf{a}_k$ contains the constant Cauchy sequence all of whose elements are a_k. Then

$$\rho^*(\mathbf{a}, \mathbf{a}_k) = \lim_{n\to\infty} \rho(a_n, a_k).$$

But since $\{a_n\}$ is a Cauchy sequence of points of M, there exists, for every positive value of ϵ, an integer $n_1(\epsilon)$ such that

$$0 \leqslant \rho(a_n, a_k) < \epsilon,$$

whenever $n > k \geqslant n_1$; hence

$$0 \leqslant \rho^*(\mathbf{a}, \mathbf{a}_k) \leqslant \epsilon,$$

whenever $k \geqslant n_1$. The point $\mathbf{a}$ of M^*, which is not a point of A^*, is the limit of a sequence of points $\{\mathbf{a}_n\}$ of A^*, and so is a point of accumulation of A^*.

It remains to show that every Cauchy sequence of points of M^* converges to a point of M^*. We prove first that every Cauchy sequence of points of A^* converges to a point of M^* and complete the proof by using the property that A^* is dense in M^*.

If $\{\mathbf{a}_n\}$ is a Cauchy sequence of points of A^* and if $\{a_n\}$ is the corresponding sequence of points of M, we have

$$\rho^*(\mathbf{a}_m, \mathbf{a}_n) = \rho(a_m, a_n)$$

and so $\{a_n\}$ is a Cauchy sequence in M. The equivalence class containing the sequence $\{a_n\}$ is a point $\mathbf{a}$ of M^*. Now if $\{b_n\}$ is a Cauchy sequence in the equivalence class $\mathbf{b}$,

$$\rho^*(\mathbf{a}, \mathbf{b}) = \lim_{m\to\infty} \rho(a_m, b_m).$$

If we take $\mathbf{b}$ to be the point $\mathbf{a}_n$ of A^*, which contains the constant sequence all of whose members are a_n, then

$$\rho^*(\mathbf{a}, \mathbf{a}_n) = \lim_{m\to\infty} \rho(a_m, a_n).$$

But since $\{a_n\}$ is a Cauchy sequence of points of M, there exists, for every positive value of ϵ, an integer $n_1(\epsilon)$ such that

$$0 \leqslant \rho(a_m, a_n) < \epsilon,$$

whenever $m > n \geqslant n_1$. It follows that

$$0 \leqslant \rho^*(\mathbf{a}, \mathbf{a}_n) \leqslant \epsilon$$

whenever $n \geqslant n_1$, and hence that the sequence $\{\mathbf{a}_n\}$ of points of A^* converges to the point $\mathbf{a}$ of M^*.

Now let $\{\mathbf{z}_n\}$ be a Cauchy sequence of points of M^*. Then, for every positive value of ϵ, there exists an integer $n_2(\epsilon)$ such that $\rho^*(\mathbf{z}_m, \mathbf{z}_n) < \epsilon$ whenever $m > n \geqslant n_2$. Keep ϵ fixed. Since A^* is dense in M^*, we can find, for every integer n, a point $\mathbf{a}_n$ of A^* such that $\rho^*(\mathbf{a}_n, \mathbf{z}_n) < \epsilon$. It follows that, if $m > n \geqslant n_2$,

$$\rho^*(\mathbf{a}_m, \mathbf{a}_n) \leqslant \rho^*(\mathbf{a}_m, \mathbf{z}_m) + \rho^*(\mathbf{z}_m, \mathbf{z}_n) + \rho^*(\mathbf{z}_n, \mathbf{a}_n) < 3\epsilon$$

so that $\{\mathbf{a}_n\}$ is a Cauchy sequence of points of A^*. By what we have proved, $\{\mathbf{a}_n\}$ converges to a point $\mathbf{a}$ of M^*. But

$$\rho^*(\mathbf{a}_m, \mathbf{z}_n) \leqslant \rho^*(\mathbf{a}_m, \mathbf{z}_m) + \rho^*(\mathbf{z}_m, \mathbf{z}_n) < 2\epsilon,$$

whenever $m > n \geqslant n_2$. Make $m \to \infty$; then

$$0 \leqslant \rho^*(\mathbf{a}, \mathbf{z}_n) \leqslant 2\epsilon,$$

whenever $n \geqslant n_2$, and so $\{\mathbf{z}_n\}$ converges to $\mathbf{a}$.

36. Some complete metric spaces

The real line with metric $|x_1 - x_2|$ is complete. So also is the complex plane with metric $|z_1 - z_2|$. For if $\{z_n\}$ is a Cauchy sequence of complex numbers, for every positive value of ϵ, there exists an integer n_0 such that $|z_m - z_n| < \epsilon$, whenever $m > n \geqslant n_0$. If we denote the real and imaginary parts of z_n by x_n and y_n, we have

$$|x_m - x_n| < \epsilon, \quad |y_m - y_n| < \epsilon,$$

whenever $m > n \geqslant n_0$. Hence $\{x_n\}$ and $\{y_n\}$ are Cauchy sequences of real numbers, which therefore converge to real numbers x and y respectively. Hence the sequence $\{z_n\}$ converges to $x + iy$.

The argument used here can be used to prove the following more general result. *If X and Y are complete metric spaces with metrics ρ_1 and ρ_2, the product space $X \times Y$ with metric*

$$\rho((x_1, y_1), (x_2, y_2)) = \sqrt{\{(\rho_1(x_1, x_2))^2 + \rho_2(y_1, y_2))^2\}}$$

is complete. From this it follows by induction that *a Euclidean space of any finite number of dimensions is complete.*

Let l^2 denote the set of all sequences $\mathbf{x} = \{x_n\}$ of complex numbers such that $\sum_1^\infty |x_n|^2$ is convergent. We have seen in §21 that

$$\rho(\mathbf{x}, \mathbf{y}) = \left\{\sum_1^\infty |x_n - y_n|^2\right\}^{\frac{1}{2}}$$

is a metric on this space. We show now that it is complete. Let $\{\mathbf{x}^k\}$ be a Cauchy sequence on l^2, where $\mathbf{x}^k = (x_1^k, x_2^k, \ldots)$. Then, for every positive value of ϵ, there exists an integer n_0 such that

$$\rho(\mathbf{x}^k, \mathbf{x}^l) = \left\{\sum_{n=1}^\infty |x_n^k - x_n^l|^2\right\}^{\frac{1}{2}} < \epsilon,$$

whenever $k > l \geqslant n_0$. It follows that, for all n,

$$|x_n^k - x_n^l| < \epsilon,$$

whenever $k > l \geqslant n_0$. But since the complex plane is complete, x_n^k tends to a limit x_n as $k \to \infty$.

Now, if $k > l \geqslant n_0$ and if m is any positive integer, we have

$$\sum_{n=1}^m |x_n^k - x_n^l|^2 < \epsilon^2.$$

If we make k tend to infinity in this finite sum, we obtain

$$\sum_{n=1}^m |x_n - x_n^l|^2 \leqslant \epsilon^2,$$

whenever $l \geqslant n_0$ and for every positive integer m. Hence the infinite series of positive terms

$$\sum_{n=1}^\infty |x_n - x_n^l|^2$$

is convergent and has a sum not exceeding ϵ^2 when $l \geqslant n_0$. But, by Minkowski's inequality

$$\left\{\sum_1^m |x_n|^2\right\}^{\frac{1}{2}} \leqslant \left\{\sum_1^m |x_n - x_n^l|^2\right\}^{\frac{1}{2}} + \left\{\sum_1^m |x_n^l|^2\right\}^{\frac{1}{2}}$$

$$\leqslant \epsilon + \left\{\sum_1^\infty |x_n^l|^2\right\}^{\frac{1}{2}}$$

so that

$$\left\{\sum_1^\infty |x_n|^2\right\}^{\frac{1}{2}}$$

is convergent. Therefore $\mathbf{x} = (x_1, x_2, x_3, \ldots)$ is a point of l^2, and

$$\rho(\mathbf{x}, \mathbf{x}^l) \leqslant \epsilon,$$

whenever $l \geqslant n_0$. The Cauchy sequence $\{\mathbf{x}^l\}$ therefore converges to the point $\mathbf{x}$ of l^2 and so l^2 is complete.

A similar argument, with the same conclusion, can be applied when the metric is

$$\rho(\mathbf{x}, \mathbf{y}) = \left\{\sum_1^\infty |x_n - y_n|^p\right\}^{1/p},$$

where $p > 1$.

37. Some complete function spaces

The simplest function space consists of all functions $x(t)$ continuous on a closed interval $[a, b]$ with metric

$$\rho(x, y) = \sup_{t \in [a, b]} |x(t) - y(t)|,$$

the space denoted by $C[a, b]$. If $\{x_n(t)\}$ is a Cauchy sequence, for every positive value of ϵ, there exists a positive integer n_0 such that

$$|x_m(t) - x_n(t)| < \epsilon,$$

whenever $m > n \geqslant n_0$, for all $t \in [a, b]$. This is the Cauchy condition for uniform convergence. But a uniformly convergent sequence of continuous functions tends to a continuous limit. Therefore the Cauchy sequence $\{x_n(t)\}$ converges to a point of $C[a, b]$, so that the space is complete.

The subspace, consisting of all polynomials defined on $[a, b]$ with the same metric, is not complete, since the limit of a uniformly convergent sequence of polynomials is not necessarily a polynomial.

The space which consists of all functions $x(t)$ which have continuous derivatives at all points of a closed interval $[a, b]$ can be made into a metric space by taking

$$\rho(x, y) = |x(a) - y(a)| + \sup_{t \in [a, b]} |x'(t) - y'(t)|.$$

This again is a complete metric space. If $\{x_n(t)\}$ is a Cauchy sequence, $\{x_n(t)\}$ converges uniformly to a continuously differentiable function $f(t)$ and $\{x_n'(t)\}$ converges uniformly to $f'(t)$.

If, however, we metrize the space of functions continuous on $[a,b]$ by taking

$$\rho(x,y) = \left\{\int_a^b |x(t)-y(t)|^2\,dt\right\}^{\frac{1}{2}},$$

the resulting metric space is not complete. To prove this, consider the sequence of continuous functions $\{x_n(t)\}$ defined on $[-1,1]$ by

$$\left.\begin{aligned} x_n(t) &= 0, & -1 \leqslant t \leqslant 0 \\ &= nt, & 0 \leqslant t \leqslant 1/n \\ &= 1, & 1/n \leqslant t \leqslant 1 \end{aligned}\right\}.$$

Then, if $m > n$,

$$\{\rho(x_m, x_n)\}^2 = (m-n)^2 \int_0^{1/m} t^2\,dt + \int_{1/m}^{1/n} (1-nt)^2\,dt$$

$$= \frac{(m-n)^2}{3m^2 n} < \frac{1}{3n} < \epsilon$$

if $n > \frac{1}{3}\epsilon^{-1}$. The sequence is thus a Cauchy sequence.

Suppose that this Cauchy sequence converged to a continuous function $x(t)$, convergence being with respect to the metric; that is, suppose that

$$\int_{-1}^{1} |x_n(t)-x(t)|^2\,dt \to 0.$$

This would imply that the integral with any limits between ± 1 also tends to zero. In particular

$$\int_{-1}^{0} |x_n(t)-x(t)|^2\,dt \to 0.$$

But $x_n(t) = 0$ when $t \leqslant 0$, and so this integral is independent of n. Therefore the continuous function $x(t)$ is such that

$$\int_{-1}^{0} |x(t)|^2\,dt = 0,$$

which gives $x(t) = 0$ when $t \leqslant 0$.

Again, if $a > 0$,

$$\int_a^1 |x_n(t)-x(t)|^2\,dt \to 0$$

as $n \to \infty$. If we choose $n > 1/a$,

$$\int_a^1 |1-x(t)|^2\,dt \to 0$$

as $n \to \infty$. As the integral is independent of n, it vanishes; and since $x(t)$ is continuous, $x(t) = 1$ for $t \geqslant a$.

But a can be as near to zero as we please. Thus we have a continuous function which vanishes when $t \leqslant 0$ and which is equal to unity when $t > 0$. Therefore the Cauchy sequence does not converge to a point of the space, and so the space is not complete.

The completion of this metric space is the space L_2 of functions whose squares are integrable in the sense of Lebesgue.

EXERCISES

1. The subspace A of a metric space M is complete. Prove that A is closed. F is a closed subset of a complete metric space M. Prove that the subspace F is complete.

2. The metric space M consists of all ordered pairs $\mathbf{x} = (x', x'')$ of real numbers with metric

$$\rho(\mathbf{x}, \mathbf{y}) = \max(|x'-y'|, |x''-y''|).$$

Prove that M is complete.

3. A space M consists of all sequences $\mathbf{x} = \{x_n\}$ of real numbers, where only a finite number of members of each sequence is non-zero. Prove that

$$\rho(\mathbf{x}, \mathbf{y}) = \left\{\sum_1^\infty (x_n - y_n)^2\right\}^{\frac{1}{2}}$$

is a metric on M.

Show that the sequence $\{\mathbf{x}_n\}$ where

$$\mathbf{x}_n = \left(1, \frac{1}{2}, \frac{1}{3}, \ldots, \frac{1}{n}, 0, 0, \ldots\right)$$

is a Cauchy sequence. Deduce that the space is not complete.

4. A space M consists of all bounded sequences $\mathbf{x} = \{x_n\}$ of real numbers. Prove that

$$\rho(\mathbf{x}, \mathbf{y}) = \sup_i |x_i - y_i|$$

is a metric on M and that M is complete.

5. A space M consists of all convergent sequences $\mathbf{x} = \{x_n\}$ of real numbers. Prove that

$$\rho(\mathbf{x}, \mathbf{y}) = \sup_i |x_i - y_i|$$

is a metric on M and that M is complete.

6. A space M consists of all sequences $\mathbf{x} = \{x_n\}$ of real numbers. Prove that

$$\rho(\mathbf{x}, \mathbf{y}) = \sum_1^\infty \frac{1}{n!} \frac{|x_n - y_n|}{1 + |x_n - y_n|}$$

is a metric on M and that M is complete.

7. $\{X_n\}$ is a sequence of complete metric spaces, the metric of X_n being ρ_n. Their Cartesian product is the set of all sequences $\mathbf{x} = \{x_n\}$ where $x_n \in X_n$. Prove that

$$\rho(\mathbf{x}, \mathbf{y}) = \sum_1^\infty \frac{1}{n!} \frac{\rho_n(x_n, y_n)}{1 + \rho_n(x_n, y_n)}$$

is a metric on the Cartesian product, and that the Cartesian product space is complete.

8. The extended complex plane $\overline{C}$ has metric

$$\rho(z, z') = \frac{2|z - z'|}{\sqrt{(1 + |z|^2)}\sqrt{(1 + |z'|^2)}},$$

$$\rho(z, \infty) = \frac{2}{\sqrt{(1 + |z|^2)}}.$$

Prove that $\overline{C}$ is complete.

9. M is the set of functions $x(t)$ continuous on every finite interval. Show that

$$\rho(x, y) = \sum_1^\infty \frac{1}{n!} \frac{\rho_n(x, y)}{1 + \rho_n(x, y)},$$

where
$$\rho_n(x, y) = \sup_{t \in [-n, n]} |x(t) - y(t)|$$

is a metric on M and that M is complete.

10. M is a complete bounded metric space. $\mathscr{F}$ is the metric space defined in Ex. 19 of Chapter 3 whose members are the closed non-empty subsets of M. Prove that $\mathscr{F}$ is complete by considering $\bigcap_{n=1}^{\infty} \overline{\bigcup_{r=n}^{\infty} F_r}$, where $\{F_n\}$ is a Cauchy sequence in $\mathscr{F}$.

11. M is the set of all analytic functions of the complex variable z, regular on the unit disc $|z| < 1$ such that

$$\sup_{0 \leqslant r < 1} \int_{-\pi}^{\pi} |f(re^{\theta i})|^2 \, d\theta < \infty.$$

Prove that $\rho(f, g) = \sup_{0 \leqslant r < 1} \left\{ \frac{1}{2\pi} \int_{-\pi}^{\pi} |f(re^{\theta i}) - g(re^{\theta i})|^2 \, d\theta \right\}^{\frac{1}{2}}$

is a metric on M.

If $|z| < r < 1$, prove that

$$|f(z)| \leqslant \frac{r}{r - |z|} \left\{ \frac{1}{2\pi} \int_{-\pi}^{\pi} |f(re^{\theta i})|^2 \, d\theta \right\}^{\frac{1}{2}}.$$

Hence show that the space M with metric ρ is complete.

12. The metric space M_1 is the set E_1 with metric ρ_1; the space M_2 is the set E_2 with metric ρ_2. The metric space $M_1 \times M_2$ is the set $E_1 \times E_2$ with metric

$$\rho(\mathbf{x}, \mathbf{y}) = \max\{\rho_1(x_1, y_1), \rho_2(x_2, y_2)\}.$$

(i) Prove that, in order that a sequence $\{\mathbf{x}^n\} = \{(x_1^n, x_2^n)\}$ in $M_1 \times M_2$ be convergent, it is necessary and sufficient that both limits

$$\lim_{n\to\infty} x_1^n = x_1, \quad \lim_{n\to\infty} x_2^n = x_2$$

should exist, and then $\lim_{n\to\infty} \mathbf{x}^n = (x_1, x_2)$.

(ii) Prove that, in order that a sequence $\{\mathbf{x}^n\}$ be a Cauchy sequence in $M_1 \times M_2$, it is necessary and sufficient that $\{x_1^n\}$ and $\{x_2^n\}$ be Cauchy sequences.

(iii) Prove that $M_1 \times M_2$ is complete if and only if M_1 and M_2 are complete.

CHAPTER 5

CONNECTED SETS

38. Separated sets

Two sets A and B in a metric space M are said to be *separated* if neither has a point in common with the closure of the other, that is, if $A \cap \bar{B} = \varnothing$, $\bar{A} \cap B = \varnothing$. If A and B are separated, they are disjoint since $A \cap B \subseteq A \cap \bar{B} = \varnothing$; but two disjoint sets are not necessarily separated.

In order that the two sets A and B may be separated, it is sufficient, but not necessary, that $d(A, B) > 0$.

For if $d(A, B) = k > 0$, the equation

$$k = \inf\{\rho(a, b) : a \in A, b \in B\}$$

implies that $\rho(a, b) \geqslant k$ for every pair of points a, b belonging to A and B respectively. If a is a point of A, it is not an adherent point of B, since the sphere $N(a; \frac{1}{2}k)$ contains no point of B. Therefore $A \cap \bar{B} = \varnothing$; similarly $\bar{A} \cap B = \varnothing$.

The condition is not necessary. Two sets can be at zero distance apart, and yet be separated. For example, on the real line, the sets $x < 0$ and $x > 0$ are separated but at zero distance; on the other hand, the sets $x < 0$, $x \geqslant 0$ are disjoint but not separated.

39. Properties of separated sets

If $A_1 \subseteq A$, $B_1 \subseteq B$ where A and B are separated, then A_1 and B_1 are separated.

Since $B_1 \subseteq B$, $\bar{B}_1 \subseteq \bar{B}$. Hence $A_1 \cap \bar{B}_1 \subseteq A_1 \cap \bar{B} \subseteq A \cap \bar{B} = \varnothing$; similarly $\bar{A}_1 \cap B_1 = \varnothing$. Hence A_1 and B_1 are separated.

Two closed sets are separated if and only if they are disjoint.

This follows at once from the fact that, if F_1 and F_2 are closed sets, $F_1 \cap \bar{F}_2 = \bar{F}_1 \cap F_2 = F_1 \cap F_2$, since $\bar{F}_1 = F_1$, $\bar{F}_2 = F_2$.

Two open sets are separated if and only if they are disjoint.

Any two separated sets are disjoint. We have to show that two disjoint open sets O_1 and O_2 are separated. If they are not separated, one of the sets $O_1 \cap \bar{O}_2$, $\bar{O}_1 \cap O_2$ is not empty. Suppose $O_1 \cap \bar{O}_2$ is not empty. Then there is a point a which belongs to O_1 and to $\bar{O}_2$. Since O_1 is open, there is a sphere $N(a; r)$ contained in O_1. But since a is an adherent point of O_2, $N(a; r)$ contains a point of O_2. Hence $O_1 \cap O_2$ is not empty, which is impossible since O_1 and O_2 are disjoint.

If the open set O is the union of two separated sets A and B, then A and B are open.

We may suppose A and B are not empty. For if $A = \varnothing$, then $B = O$, and A and B are both open.

Suppose a is any point of A. Then there exists a sphere $N(a; r)$ consisting entirely of points of O. Since a is not an adherent point of B, there exists a sphere $N(a; r_1)$, where $r_1 \leqslant r$, which contains no points of B and so consists entirely of points of A. Every point of A is thus an interior point of A; hence A is open. Similarly for B.

If the closed set F is the union of two separated sets A and B, then A and B are closed.

Since $F = A \cup B$ is closed, we have

$$A \cup B = \overline{A \cup B} = \bar{A} \cup \bar{B}.$$

Hence

$$\bar{A} = \bar{A} \cap (\bar{A} \cup \bar{B}) = \bar{A} \cap (A \cup B) = (\bar{A} \cap A) \cup (\bar{A} \cap B) = A \cup \varnothing = A$$

so that A is closed. Similarly B is closed.

40. Disconnected and connected sets

The subset A of a metric space M is said to be *disconnected* if it is the union of two non-empty separated sets. There is no special virtue in the word 'two'; for if $A = B \cup C \cup D$, where B, C, D are not empty and each pair is separated, then

$$A = B \cup (C \cup D),$$

where B and $C \cup D$ are separated. A subset of M which is not disconnected is said to be *connected*.

A metric space M is connected if and only if the only non-empty subset of M which is both open and closed is M itself.

If M is disconnected, it is the union of two separated sets B and C. Since M is closed, B and C are closed. Since B and C are the complements of C and B, B and C are also open. Hence a disconnected metric space is the union of two non-empty disjoint sets which are both open and closed.

If M is the only non-empty subset of M which is both open and closed, M is not the union of two non-empty disjoint sets which are both open and closed, and so M is not disconnected. Hence M is connected.

If M is connected, and if A is a non-empty subset which is both open and closed, A' is both closed and open. Thus M is the union of two disjoint sets A and A' which are both open and closed. Therefore A and A' are separated, which is impossible if A and A' are both non-empty. But A, by definition, is non-empty. Hence $A' = \varnothing$ and $A = M$. The only non-empty subset of M which is both open and closed is M.

A subset A of the metric space M is connected if and only if the only non-empty subset of A which is both open and closed with respect to the subspace A is A itself.

If the subset A is disconnected, $A = B \cup C$, where

$$B \cap \bar{C} = \bar{B} \cap C = \varnothing.$$

Hence $A \cap \bar{B} = (B \cup C) \cap \bar{B} = (B \cap \bar{B}) \cup (C \cap \bar{B}) = B$; similarly $A \cap \bar{C} = C$. The set B, being the intersection of A and the closed set $\bar{B}$, is closed with respect to the subspace A, and C, its complement with respect to A, is open with respect to A. Similarly C is closed and B is open with respect to A. Thus a disconnected subset A of M is the union of two non-empty disjoint subsets B and C which are both open and closed with respect to the subspace A.

The result now follows from the preceding result by regarding A as a space in its own right.

A subset A of the metric space M is disconnected if and only if $A \subseteq O_1 \cup O_2$, where O_1 and O_2 are open sets such that $A \cap O_1$ and $A \cap O_2$ are not empty, but $A \cap O_1 \cap O_2$ is empty.

This is merely another way of expressing the previous result. There is a corresponding result with O_1 and O_2 replaced by closed sets F_1 and F_2 such that $A \cap F_1$ and $A \cap F_2$ are not empty, but $A \cap F_1 \cap F_2$ is empty.

41. Properties of connected sets

B is a disconnected set; $B \subseteq O_1 \cup O_2$ where O_1 and O_2 are open sets such that $B \cap O_1 \cap O_2$ is empty, but $B \cap O_1$ and $B \cap O_2$ are not empty. If A is a connected subset of B, then either $A \subseteq O_1$ or $A \subseteq O_2$.

If A is a subset of B, $A \cap O_1 \cap O_2$ is empty. If $A \cap O_1$ and $A \cap O_2$ were both not empty, A would be disconnected. Hence one of $A \cap O_1$ and $A \cap O_2$ is empty. Now A is a subset of $O_1 \cup O_2$. If $A \cap O_2$ is empty, no point of A belongs to O_2, and so $A \subseteq O_1$. Similarly if $A \cap O_1$ is empty.

If A is a connected subset of a metric space M, and if $A \subseteq B \subseteq \bar{A}$, then B is connected. In particular, the closure of A is connected.

There is nothing to prove if $B = A$, so we suppose $A \subset B \subseteq \bar{A}$. If B is not connected, then $B \subseteq O_1 \cup O_2$, where O_1 and O_2 are certain open sets with the property that $B \cap O_1 \cap O_2$ is empty, but $B \cap O_1$ and $B \cap O_2$ are not. By the previous result, either $A \cap O_1$ or $A \cap O_2$ is empty. If $A \cap O_2$ is empty, $A \subseteq O_2'$. Hence

$$A \subset B \subseteq \bar{A} \subseteq O_2'^{-} = O_2'$$

since O_2' is closed. Therefore $B \subseteq O_2'$, and $B \cap O_2$ is empty, contrary to hypothesis. Hence B is not disconnected. Similarly if $A \cap O_1$ is empty. Hence the first result. Then take $B = \bar{A}$.

If two connected sets are not separated, their union is connected.

Let A and B be two connected sets which are not separated. If $A \cup B$ is not connected, $A \cup B = C \cup D$ where C and D are non-empty separated sets. Then

$$A = A \cap (A \cup B) = A \cap (C \cup D) = (A \cap C) \cup (A \cap D),$$

which expresses A as the union of two separated sets, which is impossible since A is connected, unless one of $A \cap C$ and $A \cap D$ is empty. Suppose that $A \cap D$ is empty, so that $A \subseteq C$. Similarly $B \subseteq C$ or $B \subseteq D$. If $B \subseteq C$, $A \cup B \subseteq C$, and hence D is empty; contrary to hypothesis. Therefore $B \subseteq D$. But this gives

$$A \cap \bar{B} \subseteq C \cap \bar{D} = \varnothing, \quad \bar{A} \cap B \subseteq \bar{C} \cap D = \varnothing,$$

which is impossible; hence $A \cup B$ is connected.

If two connected sets have a non-empty intersection, their union is connected.

Let A and B be two connected sets, whose intersection $A \cap B$ is not empty. Since $A \cap B \subseteq A \cap \bar{B}$ and $A \cap B \subseteq \bar{A} \cap B$, the result is an immediate consequence of the one we have just proved. We give here a different proof of the weaker result which is capable of extension to any family of connected sets.

Let $C = A \cup B$, and let D be any non-empty subset of C which is both open and closed with respect to the subspace C. D must intersect A or B or both; suppose it intersects A, so that $D \cap A$ is not empty. Then $D \cap A$ is a subset of the subspace A which is both open and closed with respect to the subspace A. Since A is connected, $D \cap A = A$, and hence $A \subseteq D$.

Now $A \cap B$ is not empty, and therefore $D \cap B$ is not empty. By the same argument, $B \subseteq D$. Therefore $C = A \cup B \subseteq D$. But D was a subset of C by definition; hence $D = C$.

We have thus proved that the only non-empty subset of C which is both open and closed with respect to the subspace C is C itself, and this implies that $C = A \cup B$ is connected.

If $\{A_\alpha : \alpha \in S\}$ is a family of connected sets whose intersection is not empty,

$$\bigcup_{\alpha \in S} A_\alpha$$

is connected.

Denote the union of the family of sets by C, and let D be any non-empty subset of C which is both open and closed with respect to the subspace C. The proof then follows precisely that just given.

If every pair of points in a subset A of a metric space M lies in a connected subset of A, then A is connected.

If A is disconnected, there exist open sets O_1 and O_2 of M such that $A \subseteq O_1 \cup O_2$ where $A \cap O_1$, $A \cap O_2$ are not empty, but $A \cap O_1 \cap O_2$ is empty. Let a be any point of $A \cap O_1$, b any point of $A \cap O_2$. By hypothesis, there is a connected subset B of A which contains both a and b. But $B \subseteq O_1 \cup O_2$, and $B \cap O_1$ and $B \cap O_2$ are not empty since they contain the points a and b respectively; but $B \cap O_1 \cap O_2$ is empty. This is impossible. Hence A is not disconnected.

If X is a connected space with metric ρ_1, Y a connected space with metric ρ_2 the Cartesian product $X \times Y$ with metric $\sqrt{(\rho_1^2 + \rho_2^2)}$ is connected.

The proof of this depends on the fact that a continuous mapping turns a connected set into a connected set. The result is also true if the metric on $X \times Y$ is $\rho_1+\rho_2$ or $\max(\rho_1, \rho_2)$. See p. 94.

42. Components

Let a and b be two points of a subset A of a metric space M. We write $a \sim b$ if there is a connected subset of A which contains both a and b. This relation connecting a pair of points of A is an equivalence relation. It is reflexive, since the set consisting of a single point a is connected. It is evidently symmetric. It is also transitive; for if $a \sim b$, there is a connected subset A_1 of A which contains a and b, and if $b \sim c$, there is a connected subset A_2 which contains b and c. But the set $A_1 \cup A_2$ is connected (since $A_1 \cap A_2$ is not empty) and it contains both a and c; hence $a \sim c$.

The family of equivalence classes defined by this relation is called *a partition of A into components*. Each component is a connected set, and every point of A is a member of exactly one component. The component containing a is the union of all connected subsets of A which contain a, and so is the largest connected set containing a. If A is a connected set, it has only one component, A itself. If there is more than one component, any two are separated sets.

Since the metric space M is a subset of M, we can divide M itself into components. *The components of a metric space are closed sets.*

Let C be a component of M. Since C is connected, so also is $\bar{C}$. Let a be a point of C, b a point of $\bar{C}$. Then the connected set $\bar{C}$ contains both a and b. But by definition C is the greatest connected set containing a, and so b belongs to C. Hence $\bar{C} \subseteq C$; thus C is closed.

43. Totally disconnected sets

In the Euclidean plane, every 'sphere' is a connected set. One naturally asks whether this is true in every metric space. The answer is that it is not. A simple example is provided by the set Q of rational numbers with $\rho(x,y) = |x-y|$. The sphere $N(1;1)$ with centre 1 and radius 1 in Q is the set of all rational numbers x such that $0 < x < 2$. We can partition this set into two separated sets in many ways. For instance, we can take the subsets for which $0 < x^2 < 2$ and $2 < x^2 < 4$. The sphere $N(1;1)$ is thus not connected.

If every connected subset of a set A in a metric space reduces to a single point, that is, if no two points of A lie in a connected subset of A, we say that A is *totally disconnected*. The set of rational numbers is totally disconnected; so also is the set of irrational numbers. Another example is obtained by taking any set and constructing a metric space with the discrete metric for which $\rho(x,x) = 0$ and $\rho(x,y) = 1$ when $x \neq y$.

44. Intervals on the real line

The real line is the set of real numbers with metric $|x-y|$. If a and b are two finite real numbers, the set of real numbers x such that $a \leqslant x \leqslant b$ is a bounded closed set, denoted by $[a,b]$; it is called a *closed interval*. The closed interval $[a,a]$ consists of a single point. If we omit the end points, we get $a < x < b$, which is a bounded open set, denoted by (a,b); it is called a *bounded open interval*. There are other bounded intervals which are neither open nor closed; there are $a < x \leqslant b$, denoted by $(a,b]$, and $a \leqslant x < b$ denoted by $[a,b)$.

Since there are no points $+\infty$ or $-\infty$ on the real line (though there are on the extended real line), the set $x > a$, denoted by $(a, +\infty)$ is an unbounded open interval of the real line; similarly the set $x < a$, denoted by $(-\infty, a)$ is an unbounded open interval. But the unbounded set $x \geqslant a$, denoted by $[a, +\infty)$ is closed since its complement $(-\infty, a)$ is open. And similarly for $(-\infty, a]$.

45. The connected subsets of the real line

The real line is connected.

If the real line R is disconnected, it is the union $A \cup B$ of two non-empty disjoint closed sets. Let a_1 belong to A, b_1 to B. If α is the midpoint of (a_1, b_1), α, being a point of R, belongs to A or to B. In the former case, take (α, b_1) and call it (a_2, b_2); in the latter, take (a_1, α) and call it (a_2, b_2). Continue the process indefinitely. We get a sequence of intervals $\{(a_n, b_n)\}$ with the properties:

(i) $\{a_n\}$ is a bounded increasing sequence of real numbers,
(ii) $\{b_n\}$ is a bounded decreasing sequence of real numbers,
(iii) $b_n - a_n = (b_1 - a_1)/2^{n-1}$,
(iv) for each integer n, $a_n \in A$, $b_n \in B$.

It follows that $\{a_n\}$ and $\{b_n\}$ converge to the same limit c. And since A and B are closed, c belongs to both A and B. This is impossible, since A and B are disjoint. Hence R is connected.

The non-empty connected subsets of the real line are intervals.

An interval of the form $[a, a]$ is evidently connected. The argument used for the real line can be applied to an interval of any of the types of §44, but the interval must be regarded as a subspace in the proof: closure is relative to the subspace, not to R.

Conversely, let A be a connected subset of R. Let $b = \sup A$, $a = \inf A$, where $b = +\infty$ when A is unbounded above, $a = -\infty$ when A is unbounded below. Suppose that a point ξ, where $a < \xi < b$, did not belong to A. Then $A \subseteq O_1 \cup O_2$, where O_1 and O_2 are the disjoint open sets $\{x \in R : x < \xi\}$ and $\{x \in R : x > \xi\}$; moreover $A \cap O_1$ and $A \cap O_2$ are not empty. But this implies that the subset A of R is disconnected, contrary to hypothesis. Hence, if A is connected, every point x such that

$$\inf A < x < \sup A$$

belongs to A. Thus A is an interval. The particular type of interval depends on whether either of $\inf A$ or $\sup A$ belong to A.

46. The structure of open and closed sets on the real line

Since every point of an open set on the real line is an interior point of the set, every open set is the union of a family of open intervals ('spheres') which may overlap. A more precise result is that

Every open set on the real line is the union of a finite or countable family of disjoint open intervals.

Every set can be partitioned in a unique way into separated components. The component containing a particular point is the greatest connected set containing the point. On the real line the components are intervals.

Let O be any open set on the real line. No two component intervals overlap since they are separated sets, but they may abut. If two component intervals abut at a point a, a cannot belong to either set, since otherwise the two component intervals would not be separated. Again, if a component interval with a frontier point b has no component interval abutting on it at b, b cannot belong to the component, since b, being a point of O, must be an interior point of O. Thus the components of O are separated open intervals of one of the forms (a, b), $(-\infty, b)$, (a, ∞). Since two open intervals are separated if and only if they are disjoint, every open set is the union of a family of disjoint open intervals. To complete the proof, we have to show that every family of disjoint open intervals is either finite or countable.

Let us consider first the case when the open set O is bounded; suppose that it lies in the open interval $(0, 1)$. Since this interval is of unit length, for every positive integer n, the number of components of length $\geqslant 1/n$ cannot exceed n. It follows that the number of components of length $\geqslant 1/n$ and less than $1/(n-1)$ is finite. The components of the open set O can now be arranged as a sequence in order of their length-groups, taking first those of lengths $\geqslant \frac{1}{2}$ ordered from left to right, then those that are of lengths $< \frac{1}{2}$ and $\geqslant \frac{1}{3}$ again ordered from left to right, and so on. There are only a finite number to arrange at each stage. There may only be a finite number of components in all. But if O has an infinite number of components, the fact that they can be arranged as a sequence means that they form a countable family.

We can reduce the general case to the special case when O lies in $(0, 1)$ by using the transformation $t = \frac{1}{2} \tanh x + \frac{1}{2}$. This is a continuous one-one correspondence between the points of the infinite interval $-\infty < x < \infty$ and the points of the open interval $(0, 1)$; and the open intervals of the real line correspond to open intervals of $(0, 1)$.

Every closed set on the real line is the complement of a finite or countable family of disjoint open intervals.

Since a closed set is the complement of an open set, the result is evident. Isolated points in a closed set are points where two of the disjoint open intervals abut.

EXERCISES

1. If the sets A and B are separated, and if the sets B and C are separated, prove that B and $A \cup C$ are separated.

2. If A and B are closed sets, prove that $A \cap B'$ and $B \cap A'$ are separated.

3. A is a connected subset of a connected metric space M. The complement of A is the union of two separated sets B and C. Prove that $A \cup B$ and $A \cup C$ are connected. Show also that, if A is closed, so also are $A \cup B$ and $A \cup C$.

4. A and B are two closed sets in a metric space M. Show that, if $A \cup B$ and $A \cap B$ are connected, then A is connected.

5. A and B are two sets in a metric space M and neither $A \cap B$ nor $A' \cap B$ is empty. Prove that, if B is connected, $(\mathrm{Fr}\, A) \cap B$ is not empty. Deduce that, if M is connected, every subset of M, other than M and $\varnothing$, has at least one frontier point.

6. Prove that a subset A of a metric space M is connected if and only if there exists no pair of closed sets F_1 and F_2 such that

$$A \subseteq F_1 \cup F_2, \quad F_1 \cap F_2 \cap A = \varnothing, \quad F_1 \cap A \neq \varnothing, \quad F_2 \cap A \neq \varnothing.$$

7. $C[a, b]$ is the space of all functions continuous on $a \leqslant t \leqslant b$ with metric

$$\rho(x, y) = \sup_{t \in [a, b]} |x(t) - y(t)|.$$

Prove that the space is connected.

8. If $\{A_\alpha : \alpha \in S\}$ is a family of connected sets and if one set of the family intersects all the others, prove that

$$\bigcup_{\alpha \in S} A_\alpha$$

is connected.

CHAPTER 6

COMPACTNESS

47. Uniform continuity and the Heine–Borel Theorem

Let $f(x)$ be a real function of the real variable x continuous at each point of the closed interval $[a, b]$. Then for every positive value of ϵ and for each point ξ of $[a, b]$, it is possible to find an open interval $N(\xi; \delta)$ such that $|f(x)-f(\xi)| < \epsilon$ whenever the point x of $[a, b]$ lies in $N(\xi; \delta)$; indeed for each ξ there are an infinite number of such open intervals since, if $N(\xi; \delta_1)$ is one such interval, so also is $N(\xi; \delta)$ for every $\delta < \delta_1$. The infinite family $\{N(\xi; \delta): \xi \in [a, b]\}$ of all these open intervals corresponding to all the points ξ of $[a, b]$ is called an infinite open covering of $[a, b]$; every point of $[a, b]$ belongs to at least one open interval of the family.

The Heine–Borel Theorem asserts that, from this infinite open covering of $[a, b]$, we can select a finite number of open intervals of the family, which also covers $[a, b]$. Every point of $[a, b]$ belongs to at least one of the open intervals of this finite open covering. From this follows the uniform continuity property, that, for every positive value of ϵ, there exists a positive number Δ, depending on ϵ, such that $|f(x_1)-f(x_2)| < \epsilon$ whenever the distance between the points x_1 and x_2 is less than Δ. We return to this in a more general context later.

Although the Heine–Borel Theorem arose in connexion with uniform continuity, it has nothing whatever to do with continuous functions. It merely states that, from any infinite family of open intervals of the real line which covers a bounded closed interval of the line, it is always possible to choose a finite set of intervals, a finite subcovering, which also covers the given bounded closed interval.

Another theorem concerning bounded closed intervals of the real line, closely related to the Heine–Borel Theorem, is the Bolzano–Weierstrass Theorem, which asserts that any infinite set of points contained in a closed interval $[a, b]$

possesses at least one point of accumulation which belongs to $[a, b]$.

In this chapter we are concerned with metric spaces which possess the Heine–Borel or the Bolzano–Weierstrass property.

48. Compact sets in a metric space

By an *open covering* of a metric space is meant a family of open subsets of the space with the property that every point of the space belongs to at least one of the open sets of the family. An open covering is said to be finite if it has a finite number of members; but an open covering consisting of an infinite number of open sets is said to be infinite. A metric space is said to have the Heine–Borel property or to be *compact* if from every open covering of the space it is possible to select a finite open covering. As a space with a finite number of points is evidently compact we shall assume that we are dealing with spaces and subspaces with an infinite number of points.

It is unfortunate that the terminology has not yet been standardised. Some authors say that a space which has the Heine–Borel property is 'bicompact' and reserve the term 'compact' for what we shall call 'sequentially compact'. No confusion can arise when we are dealing with metric spaces since the two properties are then equivalent.

A subset A of a metric space M is said to be compact if the subspace A is compact. A subset A of a metric space M is said to be relatively compact if the closure $\bar{A}$ of A is a compact subspace. Again, the terminology is not uniform, but this is the commonest usage. Some authors call a relatively compact set compact and call a compact set a compactum.

Every closed subset of a compact metric space is compact.

Let F be a closed subset of the metric space M. Let $\{O_\alpha: \alpha \in S\}$, where S is an infinite index set, be any infinite covering of F by sets which are open with respect to the subspace F. Then each O_α is the intersection of F and a set Ω_α open with respect to M. But since F is closed, its complement F' is open with respect to M. The family of open sets consisting of F' and $\{\Omega_\alpha: \alpha \in S\}$ is an infinite open covering of M. Since M is compact, it is possible to choose from this family a finite open covering of M consisting of F' and $\{\Omega_\alpha: \alpha \in S_1\}$ where S_1 is a finite subset of the index set S. The corresponding family $\{O_\alpha: \alpha \in S_1\}$ is a finite family of sets which are open with respect to F and cover F. Hence the subspace F has the Heine–Borel property.

49. Sequentially compact metric spaces

A metric space is said to be *sequentially compact* if every sequence of points of the space contains a convergent subsequence. A subset A of a metric space is said to be sequentially compact if the subspace A is sequentially compact: so we need only consider spaces.

A metric space is sequentially compact if and only if every infinite subset has a point of accumulation.

Let A be any infinite subset of a sequentially compact metric space M. Let $\{a_n\}$ be any sequence of distinct points of A. Since M is sequentially compact, this sequence contains a convergent subsequence $\{a_{k_n}\}$ whose limit is a, say. Then a is a point of accumulation of A.

Conversely, let M be a metric space with the property that every infinite subset has a point of accumulation. Let $\{a_n\}$ be any sequence of points of M. If any point occurs infinitely often in the sequence, the sequence contains a constant subsequence which is convergent. If not, we may assume all the members of $\{a_n\}$ are distinct. The set A of points which are the values of a_n $(n = 1, 2, 3, \ldots)$ is an infinite set, and so has at least one point of accumulation. If a is a point of accumulation of A, we can choose a sequence of points of A which converges to a, and this sequence is a subsequence of $\{a_n\}$. Hence the result that a metric space is sequentially compact if and only if every infinite subset has a point of accumulation—the Bolzano–Weierstrass property.

A compact metric space is sequentially compact.

We have to show that every infinite subset of a compact metric space has a point of accumulation. If A is an infinite subset of a compact metric space M and if A has no point of accumulation, each point of M is the centre of a sphere which contains no point of A save possibly the centre of the sphere. These spheres form an infinite open covering of M. Since M is compact, we can choose from this family of spheres, a finite number which also cover M. Since each sphere contains at most one point of A, A is a finite set, contrary to hypothesis. Hence every infinite subset of M has a point of accumulation.

A compact subset of a metric space is bounded and closed.

Let A be a subset of a metric space M, such that the subspace A is compact. Since there is nothing to prove if A is a finite set, we assume that A is an infinite set, and, by the preceding result, we may suppose it to be sequentially compact.

If A is unbounded, we can find pairs of points of A at arbitrarily large distances apart. This enables us to construct a sequence of points of A which possesses no convergent subsequence. Starting with any point a_1 of A, we choose a point a_2 such that $\rho(a_1, a_2) > 1$; we then choose a point a_3 such that

$$\rho(a_1, a_3) > 1 + \rho(a_1, a_2),$$

and so on. For every value of n, a_n is chosen so that

$$\rho(a_1, a_n) > 1 + \rho(a_1, a_{n-1}),$$

and this implies that if $m > n$,

$$\rho(a_1, a_m) > 1 + \rho(a_1, a_n).$$

Hence $$\rho(a_m, a_n) \geqslant |\rho(a_1, a_m) - \rho(a_1, a_n)| > 1.$$

From this it follows that $\{a_n\}$ contains no convergent subsequence, which is impossible. Hence A is bounded.

If a is any point of $\bar{A}$ there exists a sequence of points $\{a_n\}$ of A which converges to a; and every subsequence of $\{a_n\}$ converges to a. But since A is a compact subspace, $\{a_n\}$ has a subsequence which converges to a point of A. Hence a belongs to A, and therefore A is closed.

50. Totally bounded spaces

A metric space M is said to be *totally bounded* if, for every positive value of ϵ, M can be covered by a finite number of spheres of radius ϵ. A totally bounded space is evidently bounded.

A sequentially compact metric space is totally bounded and complete.

Let a_1 be any point of a sequentially compact metric space M. Fix $\epsilon > 0$, and choose if possible a point a_2 at a distance not less than ϵ from a_1; then choose a point a_3 at a distance not less than ϵ from a_1 and a_2; and so on. For every value of n, the distance of a_n from $a_1, a_2, \ldots, a_{n-1}$ is not less than ϵ. We show that the process must terminate. Note that, if M is bounded and ϵ is large enough, the process terminates at the first step—there is then no point at a distance not less than ϵ from a_1.

If the process did not terminate, we should have constructed a sequence $\{a_n\}$ with the property that no two points of the sequence are at a distance less than ϵ. But since M is sequentially compact, this sequence would contain a convergent subsequence, which is impossible.

For every positive value of ϵ, the process gives a finite set of points $\{a_1, a_2, \ldots, a_k\}$ such that the spheres $N(a_n; \epsilon)$, where $n = 1, 2, 3, \ldots, k$, cover M, which is therefore totally bounded. Such a finite set of points is called an ϵ-net. A sequentially compact metric space possesses an ϵ-net for every positive value of ϵ.

To prove that a sequentially compact metric space M is complete, we have to show that every Cauchy sequence converges to a point of M.

Let $\{a_n\}$ be a Cauchy sequence, so that, for every positive value of ϵ, there exists an integer k such that $\rho(a_m, a_n) < \epsilon$, whenever $m > n \geqslant k$. This Cauchy sequence contains a subsequence $\{a_{k_n}\}$ which converges to a point a of M, that is

$$\lim_{n\to\infty} \rho(a_{k_n}, a) = 0.$$

Since $\{k_n\}$ is an increasing sequence of integers, $k_m \geqslant m$. We now have

$$0 \leqslant \rho(a_n, a) \leqslant \rho(a_n, a_{k_m}) + \rho(a_{k_m}, a) < \epsilon + \rho(a_{k_m}, a)$$

whenever $m > n \geqslant k$. Make $m \to \infty$. Then

$$0 \leqslant \rho(a_n, a) \leqslant \epsilon,$$

whenever $n \geqslant k$. Therefore the Cauchy sequence $\{a_n\}$ converges to a.

It is not necessarily the case that a totally bounded metric space is sequentially compact, but

A totally bounded complete metric space is sequentially compact.

Let $\{a_n\}$ be any sequence of points in a totally bounded complete metric space M. Since M can be covered by a finite number of spheres of unit radius, at least one of these spheres will contain an infinite number of members of the sequence. Call such a sphere N_1 and let $\{a_n^1\}$ be the subsequence of $\{a_n\}$ in N_1.

But M, and therefore N_1 can be covered by a finite number of spheres of radius $\frac{1}{2}$. By the same argument, there will be a sphere N_2 of radius $\frac{1}{2}$ which contains a subsequence $\{a_n^2\}$ of $\{a_n^1\}$. The sphere N_2 can be covered by a finite number of spheres of radius $\frac{1}{3}$; and hence there will be a sphere N_3 of radius $\frac{1}{3}$ which contains a subsequence $\{a_n^3\}$ of $\{a_n^2\}$. And so on. For every positive integer k, there exists a sphere N_k of radius $1/k$ which is contained in N_{k-1} and which contains a subsequence $\{a_n^k\}$ of $\{a_n^{k-1}\}$.

Now consider the subsequence $\{a_n^n\}$ of $\{a_n\}$. If $m > k$, a_m^m belongs to N_m and therefore to N_k, so that

$$\rho(a_k^k, a_m^m) < 2/k.$$

For any positive value of ϵ, choose $k > 2/\epsilon$. Then

$$\rho(a_k^k, a_m^m) < \epsilon.$$

whenever $m > k$. The subsequence $\{a_n^n\}$ of $\{a_n\}$ is therefore a Cauchy sequence; and since M is complete, this subsequence converges to a point of M. Hence M is sequentially compact.

A totally bounded metric space is separable.

Let A_n be a set of points in the totally bounded metric space M which forms a $1/n$-net. Then since A_n is a finite set, the set of points

$$\bigcup_{n=1}^{\infty} A_n$$

is countable and is dense in M.

51. The equivalence of the two kinds of compactness

We have proved that a compact metric space is sequentially compact. We now show that a sequentially compact metric space is compact, so that it is not necessary to distinguish between the two kinds of compactness when we are dealing with metric spaces. In the proof we need the following lemma.

If the family of open sets $\{O_\alpha : \alpha \in S\}$ is an infinite open covering of the sequentially compact metric space M, there exists a positive number ϵ such that every sphere of radius ϵ is contained in at least one of the open sets O_α.

Suppose the lemma false. Then, no matter how small ϵ is, there exists a sphere of radius ϵ contained in none of the sets O_α. In particular, for every positive integer n, there exists a sphere $N(a_n; 1/n)$ which is not contained in any of the sets O_α. But since M is sequentially compact, the sequence $\{a_n\}$ of centres of this sequence of spheres contains a subsequence $\{a_{k_n}\}$ which converges to a point a of M.

Since $\{O_\alpha : \alpha \in S\}$ covers M, this point a lies in at least one of the open sets of the family; suppose it lies in O_0. The point a is an interior point of O_0, and so there exists a positive integer n_0 such that $N(a; 2/n_0)$ lies in O_0. The sphere $N(a; 1/n_0)$ contains all but a finite number of members of the subsequence $\{a_{k_n}\}$. In particular there exists a point a_{n_1} of the subsequence, where $n_1 > n_0$, contained in $N(a; 1/n_0)$. Hence

$$N(a_{n_1}; 1/n_1) \subseteq N(a_{n_1}; 1/n_0) \subseteq N(a; 2/n_0) \subseteq O_0,$$

which contradicts the definition of the sequence $\{a_n\}$. Hence the lemma is not false.

A sequentially compact metric space is compact.

Let $\{O_\alpha : \alpha \in S\}$ be any infinite open covering of a sequentially compact metric space M. If ϵ is the positive number whose existence was proved in the lemma, every sphere of radius ϵ is contained in at least one of the open sets O_α. But M, being sequentially compact, is totally bounded, so that, with this ϵ, we can find an ϵ-net $a_1, a_2, \ldots, a_k$, with the property

$$M \subseteq \bigcup_{r=1}^{k} N(a_r; \epsilon).$$

By the lemma, each sphere $N(a_r; \epsilon)$ is contained in an open set O_{α_r} of the family. Therefore

$$M \subseteq \bigcup_{r=1}^{k} O_{\alpha_r},$$

so that M has a finite open covering chosen from the infinite open covering $\{O_\alpha : \alpha \in S\}$. Hence M is compact.

To sum up, we defined a compact metric space as one which has the Heine–Borel property. We have now proved that a metric space is compact if and only if it has one of the following properties:

(i) it possesses the Bolzano–Weierstrass property;
(ii) it is sequentially compact;
(iii) it is complete and totally bounded.

52. Compactness and the finite intersection property

A quite different formulation of compactness depends on the finite intersection property. A family $\{F_\alpha : \alpha \in S\}$ of closed sets, where the index set S is infinite, is said to have the finite intersection property if, for every finite subset S_0 of S,

$$\bigcap_{\alpha \in S_0} F_\alpha$$

is not empty.

A metric space M is compact if and only if, for every infinite family $\{F_\alpha : \alpha \in S\}$ of closed sets with the finite intersection property,

$$\bigcap_{\alpha \in S} F_\alpha$$

is not empty.

Let M be compact, and let $\{F_\alpha : \alpha \in S\}$ be any family of closed sets with the finite intersection property. If $\bigcap_{\alpha \in S} F_\alpha$ is empty,

$$\bigcup_{\alpha \in S} F'_\alpha = M.$$

Hence $\{F'_\alpha : \alpha \in S\}$ is an infinite open covering of M. Since M is compact, there exists a finite subset S_0 of S such that

$$\bigcup_{\alpha \in S_0} F'_\alpha = M.$$

Taking complements,

$$\bigcap_{\alpha \in S_0} F_\alpha = \varnothing,$$

which is impossible since the family of closed sets has the finite intersection property. Hence $\bigcap_{\alpha \in S} F_\alpha$ is not empty.

Next let M be not compact. Then there exists an infinite open covering $\{O_\alpha: \alpha \in S\}$ which contains no finite open covering. For every finite subset S_0 of S, $\bigcup_{\alpha \in S_0} O_\alpha$ is a proper subset of M. Hence $\bigcap_{\alpha \in S_0} O'_\alpha$ is not empty; the family $\{O'_\alpha: \alpha \in S\}$ is a family of closed sets with the finite intersection property. But

$$\bigcup_{\alpha \in S} O_\alpha = M,$$

and so

$$\bigcap_{\alpha \in S} O'_\alpha = \varnothing.$$

Therefore if M is not compact, it is not true that every infinite family of closed sets with the finite intersection property has a non-empty intersection. Hence the result.

53. The Cartesian product of two compact spaces

It is a well-known result that a rectangle

$$a \leqslant x \leqslant b, \quad \alpha \leqslant y \leqslant \beta$$

in the Euclidean plane has the Bolzano–Weierstrass property and so is compact. This is a particular case of a more general result concerning compact metric spaces.

If X is a compact metric space with metric ρ_1 and if Y is a compact metric space with metric ρ_2, the Cartesian product space with metric $\sqrt{(\rho_1^2+\rho_2^2)}$ is compact.

We have to show that, if $\{x_n\}$ is any sequence of points in X, $\{y_n\}$ any sequence of points in Y, the sequence $\{(x_n, y_n)\}$ of points in $X \times Y$ contains a convergent subsequence.

Since X is compact, $\{x_n\}$ contains a subsequence $\{x_{k_n}\}$ which converges to a point x of X. Since Y is compact, $\{y_{k_n}\}$ contains a subsequence $\{y_{l_n}\}$ which converges to the point y of Y; and the corresponding subsequence $\{x_{l_n}\}$ of $\{x_n\}$ converges to x. Note that $l_n \geqslant k_n \geqslant n$.

For every positive value of ϵ, there exists an integer n' such that $\rho_1(x, x_{l_n}) < \epsilon/\sqrt{2}$ when $n > n'$, and an integer n'' such that $\rho_2(y, y_{l_n}) < \epsilon/\sqrt{2}$ when $n > n''$. Hence if $n_0 = \max(n', n'')$, and $n > n_0$,

$$\rho^2((x,y),(x_{l_n}, y_{l_n})) = \rho_1^2(x, x_{l_n}) + \rho_2^2(y, y_{l_n}) < \epsilon^2,$$

so that the subsequence $\{(x_{l_n}, y_{l_n})\}$ converges to (x, y). The metric space $X \times Y$ is therefore sequentially compact and hence compact.

By induction, the Cartesian product of r compact metric spaces $X_1, X_2, \ldots, X_r$ with metrics $\rho_1, \rho_2, \ldots, \rho_r$, respectively, when given a metric

$$\rho = \sqrt{(\rho_1^2 + \rho_2^2 + \ldots + \rho_r^2)},$$

is also compact. The result is also true if we take as the metric on $X_1 \times X_2 \times \ldots \times X_r$ either

$$\rho = \rho_1 + \rho_2 + \ldots + \rho_r,$$

or

$$\rho = \max(\rho_1, \rho_2, \ldots, \rho_r),$$

since the three metrics are equivalent (§61).

54. Compact sets on the real line or the Euclidean plane

The real line with metric $|x - y|$ is not compact. This is evident in several ways. First of all, the real line is not bounded, and so is not compact. Again, the family of open intervals

$$n - 1 < x < n + 1 \quad (n = 0, \pm 1, \pm 2, \ldots)$$

is an open covering of the real line; but it is impossible to select from it a finite open subcovering. But any bounded closed interval $[a, b]$ is compact, since it has the Bolzano–Weierstrass property. And since any bounded closed set of points of the real line is a closed subset of a bounded closed interval, it too is compact.

The Euclidean plane is not compact. But, by using the theorem of §53, we see that any bounded closed rectangle is compact, and this again implies that any bounded closed set of points in the Euclidean plane is compact. Similarly for a Euclidean space of any number of dimensions or for the complex plane.

55. Arzelà's Theorem

The metric space $C[a, b]$ of real functions continuous on the closed interval $[a, b]$ with metric

$$\rho(x, y) = \sup\{|x(t) - y(t)| : t \in [a, b]\}$$

is a complete metric space, but it is not compact since it is not bounded; there is no constant K such that $\rho(x, 0) \leqslant K$ for all points x of $C\,[a, b]$.

A subset M of $C[a, b]$ is compact if and only if it is closed, bounded and equi-continuous.

This result is called Arzelà's Theorem. To say that M is equi-continuous means that, for every positive value of ϵ, there exists a positive number δ such that

$$|x(t_1) - x(t_2)| < \epsilon,$$

for all points x of M, whenever $|t_1 - t_2| < \delta$; the point is that δ depends on ϵ but does not depend on the particular point x of M.

Let M be a compact subspace of $C[a, b]$. Then it is complete and totally bounded. Being totally bounded implies that M is bounded, but it also implies that it possesses an ϵ-net for every positive value of ϵ.

With any given positive value of ϵ, let $x_1, x_2, \ldots, x_k$ be an ϵ-net. Then every point x of M lies in one of the spheres $N(x_i;\ \epsilon)$ so that

$$\sup |x(t) - x_i(t)| < \epsilon$$

for at least one value of i.

Since $x_i(t)$ is continuous, it is uniformly continuous; therefore with the given value of ϵ, we choose a positive number δ_i so that

$$|x_i(t_1) - x_i(t_2)| < \epsilon,$$

whenever $|t_1 - t_2| < \delta_i$. Let $\delta = \min(\delta_1, \delta_2, \ldots, \delta_k)$.

Suppose that x lies in the sphere $N(x_j;\ \epsilon)$. Then if $|t_1 - t_2| < \delta$,

$$\begin{aligned} &|x(t_1) - x(t_2)| \\ &\quad \leqslant |x(t_1) - x_j(t_1)| + |x_j(t_1) - x_j(t_2)| + |x_j(t_2) - x(t_2)| < 3\epsilon. \end{aligned}$$

Hence the subspace M is equi-continuous.

Conversely let M be closed, bounded and equi-continuous. Then there exists a constant K such that every point of M satisfies the condition

$$\sup\{|x(t)| : t \in [a, b]\} \leqslant K;$$

and, for every positive value of ϵ, there exists a positive number δ such that

$$|x(t_1) - x(t_2)| < \epsilon,$$

whenever $|t_1 - t_2| < \delta$, for every point x of M. We prove M compact by showing that every sequence $\{x_n\}$ of points of M contains a convergent subsequence which converges to a point of M. Convergence in $C[a,b]$ is ordinary uniform convergence.

Divide $[a,b]$ into k equal parts each of length less than δ by points
$$a = t_0 < t_1 < t_2 < \ldots < t_{k-1} < t_k = b.$$
If t and t' are two points of $[t_{i-1}, t_i]$, $|x_n(t) - x_n(t')| < \epsilon$ for all n.

Consider the sequence of points $\mathbf{x}_n$ in k-dimensional Euclidean space with co-ordinates
$$\mathbf{x}_n = (x_n(\tau_1), x_n(\tau_2), \ldots, x_n(\tau_k)),$$
where τ_i is the midpoint of $[t_{i-1}, t_i]$. Since $|x_n(t)| \leqslant K$ for all n and t, the sequence is bounded and so contains a convergent subsequence $\{\mathbf{x}_{l_n}\}$ where $l_n \geqslant n$. Hence there exists a positive integer $n_0(\epsilon)$ such that
$$|x_{l_m}(\tau_i) - x_{l_n}(\tau_i)| < \epsilon$$
whenever $m > n \geqslant n_0(\epsilon)$ and $i = 1, 2, \ldots, k$. It follows that, if t is any point of $[a,b]$,
$$|x_{l_m}(t) - x_{l_n}(t)| < 3\epsilon$$
whenever $m > n \geqslant n_0(\epsilon)$.

Let $\{\epsilon_n\}$ be a decreasing sequence convergent to zero. Write $y_n^1(t)$ for $x_{ln}(t)$ with ϵ_1 for ϵ. Then, for all t in $[a,b]$,
$$|y_m^1(t) - y_n^1(t)| < 3\epsilon_1$$
whenever $m > n \geqslant n_1$, where n_1 depends on ϵ_1. From $\{y_n^1(t)\}$ we can extract a subsequence $\{y_n^2(t)\}$ such that, for all t in $[a,b]$,
$$|y_m^2(t) - y_n^2(t)| < 3\epsilon_2$$
whenever $m > n \geqslant n_2$ where n_2 depends on ϵ_2. And so on indefinitely. The sequence $\{y_n^{r-1}(t)\}$ contains a subsequence $\{y_n^r(t)\}$ such that
$$|y_m^r(t) - y_n^r(t)| < 3\epsilon_r$$
whenever $m > n \geqslant n_r$, where n_r depends only on ϵ_r.

Write $z_n(t)$ for $y_n^n(t)$. If $n > s$, $z_n(t) = y_q^s(t)$, where $q \geqslant n$; and if $m > n > s$, $z_m(t) = y_p^s(t)$ where $p > q$. Hence, if $m > n > \max(s, n_s)$, we have $|z_m(t) - z_n(t)| = |y_p^s(t) - y_q^s(t)| < 3\epsilon_s$ for all t in $[a,b]$ since $p > q > n_s$.

Now $\{\epsilon_k\}$ is a null sequence. Given any positive number ϵ, we can choose s so that $0 < 3\epsilon_s < \epsilon$. The sequence $\{z_k(t)\}$ is therefore uniformly convergent on $[a,b]$ to a continuous function $x(t)$. Since M is closed, $x(t)$ belongs to M.

EXERCISES

1. Give an example of a separable metric space which is not compact.

2. Give an example of a complete metric space which is not compact.

3. Prove that the extended complex plane with metric

$$\frac{2|z_1 - z_2|}{\sqrt{(1+|z_1|^2)}\sqrt{(1+|z_2|^2)}}$$

is compact.

4. A and B are two compact subsets of a metric space M. Prove that $A \cup B$ and $A \cap B$ are compact.

5. A and B are non-empty sets in a metric space and B is compact. Prove that $d(A, B) = 0$ if and only if $\bar{A} \cap B$ is not empty.

6. A is a compact set in a metric space M and b is a point of A'. Prove that there is a point a of A such that $\rho(a, b) = d(A, b)$.

7. A is a compact set of diameter $d(A)$. Prove that there exists a pair of points x and y of A such that $\rho(x, y) = d(A)$.

8. A and B are disjoint compact sets in a metric space M. Prove that $d(A, B) > 0$. Show also that there exist disjoint open sets O_1 and O_2 such that $A \subset O_1, B \subset O_2$.

9. Prove that a metric space M is compact if and only if it has the following property: from any infinite family of closed sets whose intersection is empty, it is possible to choose a finite family whose intersection is empty.

10. $\{F_\alpha : \alpha \in S\}$ is an infinite family of closed sets with the finite intersection property, and one set of the family is compact. Prove that $\bigcap_{\alpha \in S} F_\alpha$ is not empty.

11. Prove that, from any infinite open covering of a separable metric space one can extract a countable open covering.

12. Prove that a separable metric space is compact if, from every countable open covering, one can extract a finite open covering.

CHAPTER 7

FUNCTIONS AND MAPPINGS

56. Functions defined on an abstract set

This chapter is concerned with the extension of the idea of continuity to a function defined on an abstract set and taking values in an abstract set, possibly the same set. We recall the definition of a function or mapping given in Chapter 1.

Let E_1 and E_2 be two non-empty sets. By a mapping f of E_1 *into* E_2, we mean a relation which associates with each element a of E_1 a single well-defined element of E_2, denoted by $f(a)$ and called the image of a under the mapping f. The mapping f of E_1 into E_2 is denoted by $f: E_1 \to E_2$. The set E_1 is called the *domain* of the mapping; the set $\{f(a): a \in E_1\}$ of all points of E_2 which are images of points of E_1 is called the *range* of f. The range may be a proper subset of E_2; it may even be a single point, in which case the mapping is called a constant mapping. A function is, by definition, one-valued.

Another way of picturing a mapping is a generalization of the graph of elementary calculus. Consider the Cartesian product $E_1 \times E_2$ which is the set of all ordered pairs (x, y) where $x \in E_1$, $y \in E_2$. A mapping or function $f: E_1 \to E_2$ is a particular sort of subset of $E_1 \times E_2$; a subset with the property that, amongst the ordered pairs (x, y) which form the function, each point a of E_1 occurs once and once only. The subset f of $E_1 \times E_2$ which is a function with domain E_1 and range in E_2 then has the following properties:

(*a*) To each $x \in E_1$ there corresponds a $y \in E_2$ such that $(x, y) \in f$.

(*b*) If (x_1, y_1) and (x_2, y_2) are in f and $y_1 \neq y_2$, then $x_1 \neq x_2$.

In a mapping f of E_1 into E_2, the range may be the whole of E_2 or may be a proper subset of E_2. If the range is the whole of E_2, the mapping is said to be a *surjection* or a mapping of E_1 *onto* E_2. Every 'onto' mapping is 'into' but not all 'into' mappings are 'onto'.

If A is a subset of E_1, the set $\{f(x): x \in A\}$ is denoted by $f(A)$ and is called the image of A under the mapping f. If B is a subset of E_2, the set $\{x \in E_1 : f(x) \in B\}$ consists of all points of E_1 whose images lie in B; it is called the inverse image of B and is denoted by $f^{-1}(B)$. If no point of E_1 has an image in $B, f^{-1}(B)$ is empty. In particular, if B consists of a single point, its inverse image may be empty, may consist of one point or may consist of more than one point. This implies that f^{-1} is, in general, not a mapping or function, since a mapping or function is one-valued.

If the inverse image of each point of E_2 is either empty or consists of a single point of E_1, the mapping $f: E_1 \to E_2$ is said to be an *injection* or a *one-to-one mapping*. If $f: E_1 \to E_2$ is an injection, $f(x_1) = f(x_2)$ implies $x_1 = x_2$; and $x_1 \neq x_2$ implies $f(x_1) \neq f(x_2)$. If f is an injection, f^{-1} is a mapping of $f(E_1)$ onto E_1.

A mapping $f: E_1 \to E_2$ which is both an injection and a surjection is called a *bijection*; every point $x \in E_1$ has a unique image in E_2, every point $y \in E_2$ has a unique inverse image in E_1. An important instance of a bijection is the schlicht function of complex variable theory.

Let f be a mapping of E_1 into E_2, g a mapping of E_2 into E_3. A mapping h of E_1 into E_3 is defined by writing $h(x) = g(y)$, where $y = f(x)$ for every point x of E_1, or $h(x) = g(f(x))$. This mapping is called the *composition* or product of the mappings f and g, and is denoted by $g \circ f$.

57. Properties of mappings

In this section we list here without proofs properties of a mapping $f: E_1 \to E_2$. A, A_1, A_2 are subsets of E_1, and $\complement_{E_1} A$ denotes the complement of A with respect to E_1. B, B_1, B_2 are subsets of E_2, and $\complement_{E_2} B$ denotes the complement of B with respect to E_2.

(i) If $A_1 \subset A, f(A_1) \subseteq f(A)$.

(ii) $f(A_1 \cup A_2) = f(A_1) \cup f(A_2)$.

(iii) $f(A_1 \cap A_2) \subseteq f(A_1) \cap f(A_2)$.

(iv) $f^{-1}(B_1 \cup B_2) = f^{-1}(B_1) \cup f^{-1}(B_2)$.

(v) $f^{-1}(B_1 \cap B_2) = f^{-1}(B_1) \cap f^{-1}(B_2)$.

(vi) $f^{-1}(\complement_{E_2} B) = \complement_{E_1} f^{-1}(B)$.

(vii) If $B = f(A)$, then $A \subseteq f^{-1}(B)$.

(viii) If $A = f^{-1}(B)$, then $f(A) \subseteq B$.

The following more precise results hold only for special types of mapping.

(ix) $f(A_1) \subset f(A)$ holds for all pairs of subsets A, A_1 of E_1 such that $A_1 \subset A$ if and only if f is an injection.

(x) $f(A_1 \cap A_2) = f(A_1) \cap f(A_2)$ holds for all pairs of subsets A_1, A_2 of E_1 if and only if f is an injection.

(xi) If $B = f(A)$, then $f^{-1}(B) = A$ holds for all subsets A of E_1 if and only if f is an injection.

(xii) If $A = f^{-1}(B)$, then $f(A) = B$ holds for all subsets B of E_2 if and only if f is a surjection.

58. Mappings and sequence spaces

Of particular interest in analysis are mappings in which the domain or range or both are sequence spaces. A sequence space is a space in which every point is a sequence $\mathbf{x} = \{x_n\}$ of real or complex numbers: if the numbers are all real, the space is called a real sequence space.

The general sequence space, denoted by s, is the space of all sequences $\{x_n\}$; it can be metrized by taking

$$\rho(\mathbf{x}, \mathbf{y}) = \sum_1^\infty A_n \frac{|x_n - y_n|}{1 + |x_n - y_n|},$$

where ΣA_n is a convergent series of positive terms. The space m of bounded sequences; each point $\mathbf{x} = \{x_n\}$ is a bounded sequence, and

$$\rho(\mathbf{x}, \mathbf{y}) = \sup |x_n - y_n|$$

is the metric. The space m has a subspace c of all convergent sequences; the space c has the subspace c_0 of all null sequences. All these are complete metric spaces.

Another important sequence space is the space l^p $(p \geqslant 1)$ of sequences $\mathbf{x} = \{x_n\}$ such that $\Sigma |x_n|^p$ is convergent, with metric

$$\rho(\mathbf{x}, \mathbf{y}) = \left\{\sum_1^\infty |x_n - y_n|^p\right\}^{1/p};$$

it, too, is complete.

If A is an infinite matrix, of which the element in the nth row and kth column is $A_{n,k}$ $(n = 1, 2, 3, \ldots; k = 1, 2, 3, \ldots)$, we can define a mapping $\mathbf{y} = A(\mathbf{x})$ of a sequence space into another by

$$y_n = \sum_1^\infty A_{n,k} x_k,$$

at any rate when the infinite series are meaningful. In particular, if the matrix A is row-finite, i.e. each row of the matrix contains only a finite number of non-zero elements, the mapping maps the general sequence space s into itself since no question of convergence arises. Mappings of this type arise in the theory of summability of infinite series. For example, if

$$A_{n,k} = \frac{1}{n} \quad (k \leqslant n), \qquad A_{n,k} = 0 \quad (k > n),$$

the equation $\mathbf{y} = A\mathbf{x}$ gives

$$y_n = \frac{1}{n}(x_1 + x_2 + \ldots + x_n);$$

this is a mapping of the space c into itself. But it may happen that $\mathbf{y}$ belongs to c when $\mathbf{x}$ does not; in this case $\lim_{n\to\infty} y_n$ is called the Cesàro limit or $(C, 1)$ limit of the sequence $\{x_n\}$.

Another type of mapping maps a function space into a sequence space. The function space $C[-\pi, \pi]$ consists of all functions $x(t)$ continuous on the closed interval $[-\pi, \pi]$ of the real line with metric

$$\rho(x, y) = \sup\{|x(t) - y(t)| : t \in [-\pi, \pi]\}.$$

If we write

$$x_0 = \frac{1}{2\pi}\int_{-\pi}^{\pi} x(t)\,dt,$$

$$x_{2n-1} = \frac{1}{2\pi}\int_{-\pi}^{\pi} x(t)\,e^{-n\pi i l}\,dt \quad (n = 1, 2, 3, \ldots),$$

$$x_{2n} = \frac{1}{2\pi}\int_{-\pi}^{\pi} x(t)\,e^{n\pi i l}\,dt,$$

the function $x(t)$ is mapped onto the sequence

$$\mathbf{x} = \{x_n : n = 0, 1, 2, 3, \ldots\}$$

of Fourier coefficients. Thus we have a mapping of $C[-\pi, \pi]$ into a sequence space, e.g. into the space c_0 of null sequences, since x_n tends to zero as $n \to \infty$.

More generally, if $x(t) \in L^2[-\pi, \pi]$, the space of all functions whose squares are integrable over $[-\pi, \pi]$ in the Lebesgue sense with metric

$$\rho(x, y) = \left\{\frac{1}{2\pi}\int_{-\pi}^{\pi} |x(t) - y(t)|^2 dt\right\}^{\frac{1}{2}},$$

then, by Parseval's Theorem, the sequence $\mathbf{x} = \{x_n\}$ of Fourier coefficients belongs to l^2. In this case we have a mapping of $L^2[-\pi, \pi]$ into l^2. This mapping is a bijection; for, by the Riesz–Fischer Theorem, if $\mathbf{x} = \{x_n\}$ is any point of l^2, there exists a unique function $x(t)$ of $L^2[-\pi, \pi]$ with Fourier coefficients $\mathbf{x} = \{x_n\}$, provided we regard all functions which differ only on a set of zero measure as identical. Moreover this mapping preserves distance; if $x(t)$ and $y(t)$ are mapped onto $\mathbf{x}$ and $\mathbf{y}$ respectively,

$$\frac{1}{2\pi}\int_{-\pi}^{\pi} |x(t) - y(t)|^2 dt = \sum_{0}^{\infty} |x_n - y_n|^2.$$

This theory falls outside the scope of this book.

59. Limits and continuity

One way of introducing the ideas of limit and continuity of functions defined on an abstract space and taking values in an abstract space is to introduce the idea of distance. If we do this, we restrict our attention to metric spaces and consider the mapping of one metric space into another metric space.

Let $f: X \to Y$ be a mapping of a metric space X with metric ρ_1 into a metric space Y with metric ρ_2. If A is a subset of X, x any point of A, a any point of $\bar{A}$, we say that $f(x)$ tends to the limit b as x tends to a on A if, for every positive value of ϵ, there exists a sphere $N_1(a; \delta)$ of X such that $x \in A \cap N_1(a; \delta)$ $(x \neq a)$ implies that $f(x) \in N_2(b; \epsilon)$. The suffixes in N_1 and N_2 indicate that the spheres refer to the metrics ρ_1 and ρ_2 respectively. This is the straightforward translation into the new context of a well-known situation in the theory of functions of two variables when we consider the limit of a function $f(x, y)$ as the point (x, y) moves up to the origin along a curve A.

In the case when A is the whole metric space X, we say that $f(x)$ tends to the limit b as x tends to a if, for every positive value of ϵ, there exists a sphere $N_1(a;\delta)$ such that $x \in N_1(a;\delta)$ $(x \neq a)$ implies that $f(x) \in N_2(b;\epsilon)$. In the special case just mentioned, this reduces to the definition of the limit of a function $f(x,y)$ of two real variables as (x,y) moves up to the origin in any manner.

If $b = f(a)$, the mapping $f\colon X \to Y$ is said to be continuous at a.

The mapping $f\colon X \to Y$ of a metric space X into a metric space Y is continuous at the point a of X if and only if, for every sequence $\{x_n\}$ of points of X which converges to a, the corresponding sequence $\{f(x_n)\}$ converges to $f(a)$.

If f is continuous at a, for every positive value of ϵ there exists a sphere $N_1(a;\delta)$ such that $x \in N_1(a;\delta)$ implies that $f(x) \in N_2(f(a);\epsilon)$. But if $\{x_n\}$ is any sequence converging to a, there exists an integer n_0 such that $n \geqslant n_0$ implies that $x_n \in N_1(a;\delta)$ and hence that $f(x_n) \in N_2(f(a);\epsilon)$. Hence the sequence $\{f(x_n)\}$ converges to $f(a)$. The condition is thus necessary.

To prove the condition sufficient, suppose that it is satisfied but that f is not continuous at a. Then there exists a positive number ϵ such that, no matter how small δ may be, $x \in N_1(a;\delta)$ does not imply that $f(x) \in N_2(f(a);\epsilon)$. In this case we can find a sequence $\{x_n\}$ of points of X such that $\rho_2(f(x_n), f(a)) \geqslant \epsilon$, with

$$\rho_1(x_1, a) < 1$$

$$\rho_1(x_2, a) < \min\left(\tfrac{1}{2}, \rho_1(x_1, a)\right)$$

$$\rho_1(x_3, a) < \min\left(\tfrac{1}{3}, \rho_1(x_2, a)\right)$$

$$\cdots\cdots\cdots\cdots\cdots\cdots\cdots\cdots\cdots\cdots$$

$$\rho_1(x_n, a) < \min\left(\frac{1}{n}, \rho_1(x_{n-1}, a)\right).$$

The sequence $\{x_n\}$ converges to a; hence the sequence $\{f(x_n)\}$ converges to $f(a)$. This is impossible since $\rho_2(f(x_n), f(a)) \geqslant \epsilon$ for all n. The assumption that f is not continuous at a is thus false. Hence the result.

60. Continuous mappings

A mapping $f\colon X \to Y$ of a metric space X into a metric space Y is said to be continuous if it is continuous at each point of X.

The mapping $f: X \to Y$ of a metric space X into a metric space Y is continuous if and only if the inverse image of every open set O contained in Y is open.

If $f(X)$ is the range of f, the inverse image of an open set O contained in Y is the inverse image of $O \cap f(X)$. But $O \cap f(X)$ is open with respect to $f(X)$. There is thus no loss of generality in assuming that $f(X) = Y$, that is, that f is a mapping of X onto Y.

Let O be a non-empty open set contained in Y, and let A be its inverse image. A is not empty since every point of Y is the image of a point of X. If a is any point of A, it has a unique image b which belongs to O. Since O is open, there is a sphere $N_2(b;\epsilon)$ contained in O. Because f is continuous at a, there is a sphere $N_1(a;\delta)$ whose image is contained in $N_2(b;\epsilon)$ and therefore in O; hence $N_1(a;\delta) \subseteq A$. Thus every point a of A is the centre of a sphere contained in A, and so A is open.

Conversely suppose that the inverse image of every non-empty open set in Y is open. Let a be any point of X, and let $b = f(a)$. For every positive value of ϵ, the inverse image of the open sphere $N_2(b;\epsilon)$ is open and contains a. There is therefore a sphere $N_1(a;\delta)$ contained in the inverse image of $N_2(b;\epsilon)$. Thus for every positive value of ϵ, we can choose δ so that the image of $N_1(a;\delta)$ is contained in $N_2(b;\epsilon)$. But this is the condition that f is continuous at a. As a was any point of X, the mapping $f: X \to Y$ is continuous.

The mapping $f: X \to Y$ of a metric space X into a metric space Y is continuous if and only if the inverse image of every closed set contained in Y is closed.

This follows from the preceding result by taking complements.

61. Homeomorphisms and equivalent metrics

If the mapping $f: X \to Y$ is a bijection, $f^{-1}: Y \to X$ is also a mapping. If, in addition f is continuous on X, and f^{-1} continuous on Y, the mapping is called a *homeomorphism*. For example, in complex variable theory, the mapping

$$w = \frac{z-c}{1-\bar{c}z},$$

where $|c| < 1$, of the closed disc $|z| \leqslant 1$ onto the closed disc $|w| \leqslant 1$ is a homeomorphism.

Let X and Y be two metric spaces consisting of the same underlying set E with metrics ρ_1 and ρ_2. The metrics ρ_1 and ρ_2 are said to be *equivalent* if the identical mapping i of X onto Y, defined by $i(x) = x$ for all $x \in E$, is a homeomorphism. Since i and its inverse are continuous, open sets in Y are mapped onto open sets in X; open sets in X are mapped onto open sets in Y. In other words, the open sets of X are the same as the open sets of Y.

The family of open sets of a metric space is called its topology. If we metrize a set E with equivalent metrics, we get two metric spaces with the same topology. Properties of a metric space which depend only on its open sets and can be completely freed from the idea of distance are called topological properties. The definitions of closed set, point of accumulation, closure, interior, exterior, frontier, continuity, compactness, connectivity can all be expressed in terms of open sets. So a property of a metric space which depends only on these is a topological property.

The conditions satisfied by equivalent metrics are as follows: for every point $a \in E$ and for every positive value of ϵ, there exist positive numbers δ_1 and δ_2 such that

$$\rho_2(a, x) < \epsilon \quad \text{whenever} \quad \rho_1(a, x) < \delta_1,$$
$$\rho_1(a, x) < \epsilon \quad \text{whenever} \quad \rho_2(a, x) < \delta_2.$$

The numbers δ_1 and δ_2 depend, in general, not only on ϵ but also on a. If there exist positive numbers Δ_1, and Δ_2, depending only on ϵ and such that $\delta_1 \geqslant \Delta_1$, $\delta_2 \geqslant \Delta_2$ for all $a \in E$, the two metrics are said to be *uniformly equivalent*.

A sufficient, but not necessary, condition for two metrics to be uniformly equivalent is that there should exist positive numbers α and β such that

$$\alpha\rho_1(x, y) \leqslant \rho_2(x, y) \leqslant \beta\rho_1(x, y)$$

for all pairs of points x and y of E.

If ρ_1 is a metric on E, so also is

$$\rho_2 = \frac{\rho_1}{1 + \rho_1};$$

$\rho_1(x,y) < \epsilon$ implies that $\rho_2(x,y) < \epsilon$, and $\rho_2(x,y) < \epsilon/(1+\epsilon)$ implies that $\rho_1(x,y) < \epsilon$. Thus ρ_1 and ρ_2 are uniformly equivalent, but they do not satisfy the condition of the previous paragraph. When we are dealing with properties which can be expressed entirely in terms of open sets, with topological properties, it does not matter which metric we use. The metric ρ_2 has the advantage of being bounded.

Again, consider two metric spaces M_1 and M_2, consisting of sets E_1 and E_2 with metrics ρ_1 and ρ_2 respectively. We can metrize $E_1 \times E_2$ in various ways. If $\mathbf{x} = (x_1, x_2), \mathbf{y} = (y_1, y_2)$ are two points of $E_1 \times E_2$, the functions

$$\rho(\mathbf{x}, \mathbf{y}) = \max\{\rho_1(x_1, y_1), \rho_2(x_2, y_2)\},$$

$$\rho'(\mathbf{x}, \mathbf{y}) = \rho_1(x_1, y_1) + \rho_2(x_2, y_2),$$

$$\rho''(\mathbf{x}, \mathbf{y}) = \sqrt{\{\rho_1^2(x_1, y_1) + \rho_2^2(x_2, y_2)\}}$$

are all metrics on $E_1 \times E_2$. Since $\rho \leqslant \rho'' \leqslant \rho' \leqslant 2\rho$, they are uniformly equivalent metrics. If we are dealing with topological properties of the metrized product space $E_1 \times E_2$, it does not matter which metric we use; the first is usually the easiest to handle.

62. The continuous mapping of a connected set

If $f(x)$ is a real function of the real variable x, continuous on a bounded closed interval $[a, b]$, it is bounded on $[a, b]$, and it attains on $[a, b]$ its supremum M, its infimum m and every value between m and M. Thus $y = f(x)$ maps the bounded closed interval $[a, b]$ onto the bounded closed interval $[m, M]$. Now a bounded closed interval is (i) a connected set of the real line, (ii) a compact set on the real line; there are two corresponding generalizations, one relating to connected sets, the other to compact sets in a metric space.

If $f: X \to Y$ is a continuous mapping of a metric space X into a metric space Y, the image of every connected subset of X is connected.

Let A be a connected subset of X, and consider the continuous mapping $f: A \to f(A)$ of the subspace A onto the subspace $f(A)$ of Y. Let B be any non-empty subset of $f(A)$ which is both open and closed with respect to $f(A)$. Since f is continuous, $f^{-1}(B)$ is a non-empty subset of the subspace A which is both open and closed with respect to A. But, since A is connected, the only non-empty subset which is both open and closed with respect to A is A itself, and so $f^{-1}(B) = A$. The mapping $f: A \to f(A)$ is a surjection, and so $f(A) = B$. Therefore the only non-empty subset of the subspace $f(A)$ which is both open and closed with respect to $f(A)$ is the whole subspace $f(A)$; hence $f(A)$ is connected.

As an application of this result, we prove the following theorem concerning product spaces.

Let the set E_1 with metric ρ_1 be a connected metric space M_1; let the set E_2 with metric ρ_2 be a connected metric space M_2. The metric space $M_1 \times M_2$, consisting of the set $E_1 \times E_2$ with metric

$$\rho = \max(\rho_1, \rho_2)$$

is connected.

Let (a_1, a_2), (b_1, b_2) be any two points of $M_1 \times M_2$. The mapping $f: M_1 \to M_1 \times M_2$, defined by $f(x_1) = (x_1, b_2)$ maps M_1 onto the subset $M_1 \times \{b_2\}$ of $M_1 \times M_2$. The mapping is continuous; for, if x_1 and y_1 are any two points of M_1,

$$\rho(f(x_1), f(y_1)) = \max(\rho_1(x_1, y_1), \rho_2(b_2, b_2)) = \rho_1(x_1, y_1).$$

The mapping conserves distance; it is isometric.

Since M_1 is connected, its image

$$f(M_1) = M_1 \times \{b_2\}$$

is connected. Similarly $\{a_1\} \times M_2$ is connected. But these two subsets of $M_1 \times M_2$ have the point (a_1, b_2) in common. Hence the set

$$E_{a_1, b_2} = M_1 \times \{b_2\} \cup \{a_1\} \times M_2$$

is connected.

The set E_{a_1, b_2} contains the points (a_1, a_2) and (b_1, b_2). Therefore the connected component of $M_1 \times M_2$ which contains (a_1, a_2) contains any other point (b_1, b_2). Hence the component which contains (a_1, a_2) is the whole space $M_1 \times M_2$, and so $M_1 \times M_2$ is connected.

It is evident that the proof would be unaltered if we had used instead of ρ either of the uniformly equivalent metrics ρ', ρ'' of § 61.

63. Connected sets in the Euclidean plane

The theorem of §62 enables us to prove the following result which characterizes connected open sets in the Euclidean plane.

A non-empty open set in the Euclidean plane is connected if and only if every pair of its points can be joined by a polygonal arc which lies in the set.

It is convenient to represent the point of the plane with rectangular cartesian co-ordinates (x, y) by the complex number $z = x+iy$, so that the metric is $\rho(z_1, z_2) = |z_1 - z_2|$. The sphere $N(a; r)$ is the disc $|z-a| < r$ with centre a and radius r. A polygonal arc is a finite chain of straight segments.

To prove the condition necessary, let O be any non-empty open connected set, and let a be any point of O. Let O_1 be the subset of O consisting of points which can be joined to a by a polygonal arc; let O_2 be the complement of O_1 with respect to O, the set of points which cannot be joined to a by a polygonal arc. O_1 and O_2 are disjoint sets; we show that they are open sets.

If a_1 is any point of O_1, there is a disc $N(a_1; \epsilon_1)$ contained in O since O is open; every point of this disc can be joined to a_1 by a segment lying in O and can therefore be joined to a by a polygonal arc. Every point of the disc belongs to O_1, and hence a_1 is an interior point of O_1. As a_1 was any point of O_1, O_1 is open.

Next let a_2 be any point of O_2. Since O is open, there is a disc $N(a_2; \epsilon_2)$ contained in O. If any point a_3 of this disc belonged to O_1, we could join a_2 to a by a polygonal arc consisting of the polygonal arc from a to a_3, followed by the segment from a_3 to a_2. Thus a_2 would belong to O_1 which is impossible since O_1 and O_2 are disjoint. Since every point of O_2 is an interior point of O_2, O_2 is open. We have thus expressed the open connected set O as the union of two disjoint open sets O_1, O_2 which is impossible unless one of the sets O_1, O_2 is empty. Since O_1 is not empty, O_2 is empty; and every point of O can be joined to any point of O by a polygonal arc lying in O.

To prove the condition sufficient, let O be an open set, every pair of whose points can be joined by a polygonal arc lying in O. If O is not connected, it is the union of two disjoint open sets O_1

and O_2. Let a_1 be any point of O_1, a_2 any point of O_2. Since these points can be joined by a polygonal arc lying in O, one link of the chain must be a segment with ends b_1 and b_2 belonging to O_1 and O_2 respectively; call this open segment L, its closure $\bar{L}$. Then, on $\bar{L}$, $z = f(t)$, where

$$f(t) = b_1(1-t) + b_2 t \quad (0 \leqslant t \leqslant 1).$$

Evidently f is a continuous mapping of the open interval $(0, 1)$ onto L, and a continuous mapping of the closed interval $[0, 1]$ onto $\bar{L}$.

Consider the subsets I_1, I_2 of $(0, 1)$ defined by

$$I_1 = \{t : f(t) \in O_1\}, \quad I_2 = \{t : f(t) \in O_2\}.$$

We have $$I_1 \cup I_2 = (0, 1), \quad I_1 \cap I_2 = \varnothing$$

since O_1 and O_2 are disjoint.

Now $O_1 \cap L$ and $O_2 \cap L$ are open with respect to L. Hence their inverse images I_1 and I_2 are open with respect to $(0, 1)$. We have thus expressed the open interval $(0, 1)$ as the union of two disjoint open sets, which is impossible since $(0, 1)$ is connected, unless one of I_1 and I_2 is empty.

If I_1 is empty, $O_1 \cap \bar{L}$ consists of the single point $\{b_1\}$, and its inverse image consists of the single point $t = 0$. But $O_1 \cap \bar{L}$ is open with respect to $\bar{L}$, and therefore its inverse image is open with respect to $[0, 1]$, which is impossible. Hence I_1 is not empty; similarly I_2 is not empty. The assumption that O is not connected is thus proved to be false.

This proof will also apply with the obvious modifications to a Euclidean space of any finite number of dimensions.

A similar theorem is:

A non-empty open set in the Euclidean plane is connected if and only if every pair of its points can be joined by an arc which lies in the set.

An arc is defined to be a set of points which is the continuous image of a bounded closed interval of the real line.

64. Continuous mappings of a compact metric space

If $f: X \to Y$ is a continuous mapping of a compact metric space X into a metric space Y, it is uniformly continuous.

To say that f is uniformly continuous means that, for every positive value of ϵ, there exists a positive number Δ such that, if x_1 and x_2 are any two points of X at a distance apart less than Δ, the distance between their images $f(x_1)$ and $f(x_2)$ is less than ϵ. The proof follows the corresponding proof for real functions of a real variable.

Let a be any point of X. Then, for every positive value of ϵ, we can choose a positive number $\delta(\epsilon, a)$ such that $\rho_1(x, a) < \delta(\epsilon, a)$ implies $\rho_2(f(x), f(a)) < \epsilon$. In fact there are an infinite number of such numbers $\delta(\epsilon, a)$; for if we find one such, any smaller positive number will do just as well.

The family of spheres

$$\{N_1(a; \tfrac{1}{2}\delta(\epsilon, a)) : a \in X\}$$

is an open covering of X. Since X is compact, we can choose from this covering a finite subcovering

$$\{N_1(a_r; \tfrac{1}{2}\delta_r) : r = 1, 2, \ldots, k\},$$

where $a_1, a_2, \ldots, a_k$ are points of X and $\delta_r = \delta(\epsilon, a_r)$. Let $\delta(\epsilon)$ be the least of the numbers $\delta_1, \delta_2, \ldots, \delta_k$.

If x_1 is any point of X, it lies in at least one sphere of the finite subcovering, say in $N_1(a_s; \frac{1}{2}\delta_s)$, so that

$$\rho_2(f(x_1), f(a_s)) < \epsilon.$$

If x_2 is any point such that $\rho_1(x_1, x_2) < \frac{1}{2}\delta(\epsilon)$, x_2 lies in the sphere $N_1(a_s; \delta_s)$, and hence

$$\rho_2(f(x_2), f(a_s)) < \epsilon.$$

Therefore
$$\rho_2(f(x_1), f(x_2)) < 2\epsilon,$$

whenever
$$\rho_1(x_1, x_2) < \tfrac{1}{2}\delta(\epsilon).$$

Hence if $\Delta(\epsilon) = \frac{1}{2}\delta(\frac{1}{2}\epsilon)$, $\rho_1(x_1, x_2) < \Delta(\epsilon)$ implies that

$$\rho_2(f(x_1), f(x_2)) < \epsilon.$$

If f: $X \to Y$ is a continuous mapping of a compact metric space X onto a metric space Y, Y is compact.

Let $\{y_n\}$ be any sequence of points of Y. For each y_n, we can choose a point x_n of X such that $f(x_n) = y_n$, since the mapping is a surjection. But X is compact, and so the sequence $\{x_n\}$ contains a convergent subsequence $\{x_{k_n}\}$; the continuity of f implies that the corresponding subsequence $\{y_{k_n}\}$ is convergent. Thus every sequence of points of Y contains a convergent subsequence. This proves that Y is compact.

If f: $X \to Y$ is a continuous mapping of a metric space X into a metric space Y, the image of any compact subset of X is compact.

When we say that A is a compact subset of X we mean that the subspace consisting of the points of A with induced metric is a compact space. Thus we are concerned with the continuous mapping of a compact space A onto a space $f(A)$. By the preceding result, $f(A)$ is compact.

If f: $X \to Y$ is a continuous mapping of a compact metric space X into a metric space Y, the image of any closed set in X is a closed set in Y.

Since X is compact, a set F closed in X is compact. Its image $f(F)$ is therefore compact and so is bounded and closed.

If f: $X \to Y$ is a continuous bijection of a compact metric X space onto a metric space Y, then f is a homeomorphism.

All we have to prove is that f^{-1}: $Y \to X$ is continuous. This is true since $f = (f^{-1})^{-1}$ maps every closed set in X onto a closed set in Y.

65. Extension theorems

Let X and Y be abstract sets, and let A be a proper subset of X. If f is a mapping of A into Y, a mapping g: $X \to Y$ was called an extension of f if $f(x) = g(x)$ for every point x of A; and f is then the restriction of g to A.

If f and g are two continuous mappings of a metric space X into a metric space Y, the set of points x belonging to X such that $f(x) = g(x)$ is closed.

It suffices to show that the set of points A at which $f(x) \neq g(x)$ is open. Let $a \in A$, and let $\rho_2(f(a), g(a)) = 3k$; by hypothesis $k > 0$. Since f and g are continuous, for every positive value of ϵ, there exists a positive number $\delta(\epsilon)$ such that

$$\rho_2(f(x), f(a)) < \epsilon, \quad \rho_2(g(x), g(a)) < \epsilon$$

whenever $$\rho_1(x, a) < \delta(\epsilon),$$

and therefore

$$3k = \rho_2(f(a), g(a)) \leqslant \rho_2(f(a), f(x)) + \rho_2(f(x), g(x)) + \rho_2(g(x), g(a))$$
$$< \rho_2(f(x), g(x)) + 2\epsilon.$$

Take $\epsilon = k$; then, whenever $\rho_1(x, a) < \delta(k)$, $\rho_2(f(x), g(x)) > k$. Hence the sphere $N_1(a; \delta(k))$ belongs to A. Every point of A is therefore an interior point of A, and so A is open.

If f and g are two continuous mappings of a metric space X into a metric space Y, and if $f(x) = g(x)$ at all points of a set everywhere dense in X, then $f = g$.

We know that the set F of points at which $f(x) = g(x)$ is closed. If $f(x) = g(x)$ on a set A everywhere dense in X, $A \subseteq F$, and therefore $\bar{A} \subseteq \bar{F} = F$. But $\bar{A} = X$; hence $F = X$, and so $f = g$ everywhere on X. Thus a continuous mapping $f\colon X \to Y$ is determined uniquely by the values f takes on a set everywhere dense in X.

If a is an adherent point of a subset A of a metric space X, we said that $f(x)$ tends to the limit l as x tends to a on A if, for every positive value of ϵ, there exists a sphere $N_1(a; \delta)$ such that $f(x) \in N_2(l; \epsilon)$ for all points $x \neq a$ belonging to $A \cap N_1(a; \delta)$. We then write

$$\lim_{x \to a,\, x \in A} f(x) = l.$$

If $a \in A$ and $l = f(a)$, we say that $f(x)$ *is continuous at a on A.*

Let A be a set everywhere dense in a metric space X, and let $f\colon A \to Y$ be a mapping of A into a metric space Y. In order that there should exist a continuous mapping $g\colon X \to Y$ coinciding with f on A, it is necessary and sufficient that, for every point a of X, $f(x)$

should tend to a limit as x tends to a on A, and that, if a is a point of A, the limit is $f(a)$. The extension g of f from A to X is unique.

The uniqueness follows from the previous result. We have to prove the conditions necessary and sufficient.

The conditions are necessary. Suppose that $f(x)$ has a continuous extension $g(x)$ such that $g(x) = f(x)$ on A. Then, if a is a point of A,

$$f(a) = g(a) = \lim_{x \to a} g(x) = \lim_{x \to a,\, x \in A} g(x) = \lim_{x \to a,\, x \in A} f(x)$$

so that $f(x)$ tends to $f(a)$ as $x \to a$ on A.

If a is not a point of A, it is a point of $\bar{A}$, since A is everywhere dense in X. Then

$$g(a) = \lim_{x \to a} g(x) = \lim_{x \to a,\, x \in A} g(x) = \lim_{x \to a,\, a \in A} f(x),$$

so that $f(x)$ tends to a limit as x tends to a on A.

The conditions are also sufficient. For every point a of X, we define $g(a)$ by

$$g(a) = \lim_{x \to a,\, a \in A} f(x).$$

If $a \in A$, this limit is $f(a)$, so that $g = f$ on A.

To show that g is continuous, we observe that, by the definition of a limit, for every positive value of ϵ, there exists a positive number δ such that

$$f(x) \in N_2(g(a); \tfrac{1}{2}\epsilon),$$

whenever $\qquad x \neq a \quad \text{and} \quad x \in A \cap N_1(a; \delta).$

If b is any point of $N_1(a; \delta)$, $g(b)$ is the limit of $f(x)$ as $x \to b$ on $A \cap N_1(a; \delta)$. Therefore

$$g(b) \in \overline{f(A \cap N_1(a; \delta))} \subseteq \overline{N_2(g(a); \tfrac{1}{2}\epsilon)} \subset N_2(g(a); \epsilon),$$

so that $g(x)$ is continuous at a. Hence g is continuous on X. Thus g is the unique continuous extension of f.

Let f be a uniformly continuous mapping of a set A, everywhere dense in the metric space X, into a complete metric space Y. Then there exists a unique continuous mapping $g\colon X \to Y$, coinciding with f on A, and g is uniformly continuous.

By the previous result, to prove the existence of g, we have to show that, for every point $a \in X, f(x)$ tends to a limit as $x \to a$ on A.

Let a be any point of X. Let $\{x_n\}$ be any sequence of points of A which converges to a. Since f is uniformly continuous on A, for every positive value of ϵ, there exists a positive number $\delta(\epsilon)$ such that $\rho_2(f(x), f(x')) < \epsilon$ whenever x and x' belong to A and $\rho_1(x, x') < \delta$. Since $\{x_n\}$ is a convergent sequence, we can choose a positive integer n_0 such that $\rho_1(x_m, x_n) < \delta$, whenever $m > n \geqslant n_0$; hence $\rho_2(f(x_m), f(x_n)) < \epsilon$, whenever $m > n \geqslant n_0$. This shows that $\{f(x_n)\}$ is a Cauchy sequence in the complete metric space Y and hence it converges to a limit which we denote by $g(a)$.

We prove that $f(x)$ converges to $g(a)$ as $x \to a$ on A. Let x be any point of A such that $0 < \rho_1(x, a) < \frac{1}{2}\delta$. Since $\{x_n\}$ converges to a, we can choose an integer n_1 so that $\rho_1(a, x_n) < \frac{1}{2}\delta$, whenever $n \geqslant n_1$. Hence $\rho_1(x, x_n) < \delta$ and therefore $\rho_2(f(x), f(x_n)) < \epsilon$ whenever $n \geqslant n_1$. Make $n \to \infty$, remembering that $\rho_2(b, y)$ is a continuous function of y. We then have $\rho_2(f(x), g(a)) \leqslant \epsilon$, whenever $x \in A$ and $\rho_1(x, a) < \frac{1}{2}\delta$. This proves that $f(x)$ tends to the limit $g(a)$ as $x \to a$ on A.

To prove that g is uniformly continuous, let a and a' be two points of X. Let $\{x_n\}$ be a sequence of points of A converging to a; let $\{x'_n\}$ be a sequence of points of A converging to a'. Suppose that $\rho_1(a, a') < \frac{1}{3}\delta$. We can choose a positive integer n_2 such that $\rho_1(x_n, a) < \frac{1}{3}\delta, \rho_1(x'_n, a') < \frac{1}{3}\delta$, when $n \geqslant n_2$. Since

$$\rho_1(x_m, x'_n) \leqslant \rho_1(x_m, a) + \rho_1(a, a') + \rho_1(a', x'_n) < \delta$$

whenever $m > n \geqslant n_2$ it follows that $\rho_2(f(x_m), f(x'_n)) < \epsilon$. Make $m \to \infty$ and then $n \to \infty$. This gives $\rho_2(g(a), g(a')) \leqslant \epsilon$, whenever $\rho_1(a, a') < \frac{1}{3}\delta$, so that g is uniformly continuous.

If we omit the condition that the continuity of the mapping of the set A, everywhere dense in the metric space X, into Y is uniform, the conclusion that f has a continuous extension to X does not necessarily follow. A simple example is as follows. Let A be the set $0 < |x| \leqslant 1$ on the real line: it is everywhere dense in $[-1, 1]$. Let $f(x) = \cos(\pi/x)$ on A. Then $f(x)$ is continuous on A; but it is not uniformly continuous since we can find pairs of points x and y of A, as near as we please, for which $f(x)-f(y) = 2$; for example, take $x = 1/(2n)$, $y = 1/(2n+1)$. This function does not have a continuous extension to $[-1, 1]$.

66. Uniform convergence

Let $\{f_n(x)\}$ be a sequence of functions defined on a metric space X and taking values in a metric space Y. If the sequence converges at each point of X, it is said to be point-wise convergent; $\lim_{n\to\infty} f_n(x)$ is then a function defined at every point of X. The condition for the sequence $\{f_n(x)\}$ to be point-wise convergent to a function $f(x)$ is that, for every positive value of ϵ and for every point x of X, there should exist an integer $n_0(\epsilon, x)$ such that

$$\rho_2(f_n(x), f(x)) < \epsilon,$$

whenever $n \geqslant n_0(\epsilon, x)$. We make $n_0(\epsilon, x)$ definite by choosing it as small as possible.

The integer $n_0(\epsilon, x)$ is a real function defined on X, but it is not necessarily bounded. If it is bounded, we say that the sequence converges uniformly to $f(x)$ on X, just as we do in the case of the theory of functions of a real or complex variable. The condition for this is that

$$\sup_{x\in X} n_0(\epsilon, x) = n_1(\epsilon) < \infty.$$

Thus the sequence $\{f_n(x)\}$ converges uniformly to $f(x)$ on X, if, for every positive value of ϵ, there exists a positive integer $n_1(\epsilon)$ such that

$$\rho_2(f_n(x), f(x)) < \epsilon,$$

whenever $n \geqslant n_1(\epsilon)$, for all points x of X.

The family of bounded functions, defined on a metric space X and taking values in a metric space Y, can be regarded as an abstract space Z, and we can define a metric on Z by writing

$$\rho_3(f, g) = \sup\{\rho_2(f(x), g(x)) : x \in X\}.$$

It is easily shown that this function does satisfy the conditions for a metric. This leads to an alternative definition of uniform convergence.

If

$$M_n = \rho_3(f_n, f),$$

the sequence $\{f_n(x)\}$ *converges uniformly to* $f(x)$ *on* X *if and only if* $M_n \to 0$ *as* $n \to \infty$.

Let $M_n \to 0$. Then, for every positive value of ϵ, there exists an integer $n_1(\epsilon)$ such that $0 \leqslant M_n < \epsilon$ whenever $n \geqslant n_1$. Hence

$$\rho_3(f_n, f) < \epsilon,$$

whenever $n \geqslant n_1$. But this implies that

$$\rho_2(f_n(x), f(x)) < \epsilon,$$

whenever $n \geqslant n_1$, for all points x of X, so that the sequence converges uniformly to $f(x)$.

Conversely, let the convergence be uniform. Then, for every positive value of ϵ, there exists an integer $n_1(\epsilon)$ such that

$$\rho_2(f_n(x), f(x)) < \tfrac{1}{2}\epsilon,$$

whenever $n \geqslant n_1(\epsilon)$ for all points x of X. Hence

$$M_n = \rho_3(f_n, f) = \sup\{\rho_2(f_n(x), f(x)) : x \in X\} \leqslant \tfrac{1}{2}\epsilon < \epsilon,$$

whenever $n \geqslant n_1(\epsilon)$. Therefore $M_n \to 0$ as $n \to \infty$.

In other words, uniform convergence on X is equivalent to ordinary convergence on Z.

67. Uniform convergence and continuity

The limit of a uniformly convergent sequence of continuous functions of a real variable is continuous. The same is true in a metric space.

If $\{f_n(x)\}$ *is a sequence of continuous functions defined on a metric space* X *and taking values in a metric space* Y, *and if the sequence converges uniformly to* $f(x)$ *on* X, *then* $f(x)$ *is continuous.*

If the convergence is uniform, then, for every positive value of ϵ, we can find an integer n_0 such that $\rho_2(f_n(x), f(x)) < \epsilon$ whenever $n \geqslant n_0$ for all x belonging to X. Let m be any fixed integer $\geqslant n_0$. Then $f_m(x)$ is continuous. Hence if a is any fixed point of X, we can find a positive number δ such that

$$\rho_2(f_m(x), f_m(a)) < \epsilon \quad \text{whenever} \quad x \in N_1(a; \delta).$$

Therefore

$$\rho_2(f(x),f(a)) \leqslant \rho_2(f(x),f_m(x)) + \rho_2(f_m(x),f_m(a)) + \rho_2(f_m(a),f(a)) < 3\epsilon,$$

where $x \in N_1(a;\ \delta)$, and so $f(x)$ is continuous at a. As a was any point of X, the result follows.

68. Real functions on a metric space

A real function on a metric space X is a mapping $f: X \to R$ of X into the real line. A simple example of such a function is $\rho_1(x, a)$ where a is any fixed point of X. The function $\rho_1(x, a)$ is continuous on X: in fact. it is uniformly continous. For if x_1 and x_2 are two points of X,

$$|\rho_1(x_1, a) - \rho_1(x_2, a)| \leqslant \rho_1(x_1, x_2) < \epsilon,$$

whenever $\rho_1(x_1, x_2) < \epsilon$. More generally, if A is a fixed non-empty subset of X, $d(x, A)$ is a uniformly continuous real function on X.

In the simplest case, the metric space X is the real line, and the results of this section then reduce to known results in the theory of functions of a real variable.

If f is a real function on a metric space X and continuous on a compact subset A of X, f is uniformly continuous and bounded on A and attains its supremum and infimum on A.

Since the subspace A is compact, uniform continuity follows from a previous result. The image $f(A)$ is a compact set on the real line; it is therefore bounded and closed, and so contains its supremum and infimum.

If f is a real function on a metric space X and continuous on a connected subset A of X, the image $f(A)$ is an interval.

The image $f(A)$ is a connected set of the real line and so is an interval. If a and b are any two points of $f(A)$, every point between a and b belongs to $f(A)$.

If f and g are two continuous mappings of a metric space X into the real line, the set of points at which $f(x) > g(x)$ is open.

Let A be the set of points of X at which $f(x) > g(x)$. If A is empty, it is open. If A is not empty let a be any point of A, so that $f(a) > g(a)$. Choose a real number b so that $f(a) > b > g(a)$.

Since $I_1 = (b, +\infty)$ is an open set, the inverse image $f^{-1}(I_1)$ is an open set containing a; on it, $f(x) > b$. Similarly if $I_2 = (-\infty, b)$, the inverse image $g^{-1}(I_2)$ is an open set containing a on which $g(x) < b$. It follows that

$$f^{-1}(I_1) \cap g^{-1}(I_2)$$

is an open set containing a on which $f(x) > b > g(x)$. But evidently

$$f^{-1}(I_1) \cap g^{-1}(I_2) \subseteq A.$$

Thus a, being a point of an open subset of A, is an interior point of A. Hence A is open.

It follows that the set of points on which $f(x) \leqslant g(x)$, being the complement of A, is closed.

If f and g are two continuous mappings of a metric space X into the real line, and if $f(x) \leqslant g(x)$ on a set A everywhere dense in X, then $f(x) \leqslant g(x)$ on X.

We know that $f(x) \leqslant g(x)$ on a closed set $F \subseteq X$. This implies that $A \subseteq F$. Hence $X = \bar{A} \subseteq \bar{F} = F$, which proves the result.

If $\{f_n\}$ is a monotonic sequence of real functions, each continuous on a compact metric space X, and if the sequence converges to a continuous function f at each point of X, the convergence is uniform on X. (Dini's Theorem.)

Suppose that the sequence is increasing. For every positive value of ϵ and for each point a of X, there exists an integer n such that $0 \leqslant f(a) - f_n(a) < \epsilon$; n depends on ϵ and on a. But since f and f_n are continuous, there exists a sphere $N(a;\, \delta)$ such that

$$|f(x) - f(a)| < \epsilon, \quad |f_n(x) - f_n(a)| < \epsilon,$$

whenever $x \in N(a;\, \delta)$; the radius δ depends on a, ϵ and n and therefore on a and ϵ. It follows that $0 \leqslant f(x) - f_n(x) < 3\epsilon$, whenever $x \in N(a;\, \delta)$.

There is such a sphere for each point a of X, and the set of all the spheres forms an open covering of X. But since X is compact, we can choose a finite set of points $a_1, a_2, \ldots, a_k$ such that the corresponding spheres form a finite open covering of X. Let n_r and δ_r be the values of n and δ corresponding to the point a_r.

If x is any point of X, it belongs to at least one sphere of this finite family. Suppose it belongs to the rth Then if $n \geqslant n_r$,

$$0 \leqslant f(x) - f_n(x) \leqslant f(x) - f_{n_r}(x) < 3\epsilon.$$

Let n_0 be the greatest of $n_1, n_2, \ldots, n_k$. It follows that for every positive value of ϵ and for all $x \in X$,

$$0 \leqslant f(x) - f_n(x) < 3\epsilon,$$

whenever $n \geqslant n_0$. The convergence is therefore uniform.

69. The extension of real continuous functions

In this section we prove the Extension Theorem of Tietze and Urysohn, that *if f is a bounded continuous real function defined on a closed subset F of a metric space X, there exists a continuous real function g defined on X which coincides with f on F.*

The proof depends on two lemmas.

LEMMA 1. *For every pair of disjoint closed subsets A and B of a metric space X, there exists a continuous real function g such that $g(x) = -1$ for $x \in A$, $g(x) = +1$ for $x \in B$, $-1 < g(x) < 1$ for $x \in A' \cap B'$.*

The space X is the union of A, B and $A' \cap B'$. The function

$$g(x) = \frac{d(x, A) - d(x, B)}{d(x, A) + d(x, B)}$$

is continuous on X and has the desired properties.

LEMMA 2. *If f is a continuous real function defined on the closed subset F of a metric space X such that $|f(x)| \leqslant 3$, there exists a continuous real function g defined on X such that $|g(x)| \leqslant 1$ for all $x \in F$, $|f(x) - g(x)| \leqslant 2$ for all $x \in F$, and $|g(x)| < 1$ for all $x \in F'$.*

Let A be the set of points of F such that $-3 \leqslant f(x) \leqslant -1$. Then A is a closed subset of the closed subspace F, and so is a closed subset of X. Similarly if B is the set of points of F such that $1 \leqslant f(x) \leqslant 3$, B is closed. Moreover A and B are disjoint. The function $g(x)$ of Lemma 1 satisfies the stated conditions.

There is no merit in the constant 3. If $|f(x)| \leqslant M$, we can find $g(x)$ so that

$$|g(x)| \leqslant \tfrac{1}{3}M \text{ on } F, \ |f(x) - g(x)| \leqslant \tfrac{2}{3}M \text{ on } F, \ |g(x)| < \tfrac{1}{3}M \text{ on } F'$$

merely by multiplication everywhere by the factor $\frac{1}{3}M$.

To prove the extension theorem, we construct a Cauchy

sequence $\{g_n(x)\}$ of real functions, continuous on X, whose limit has the desired property.

Let $|f(x)| \leqslant M$ on F. By Lemma 2, we can find a real function $g_1(x)$, continuous on X, such that

$$|f(x)-g_1(x)| \leqslant \tfrac{2}{3}M \quad (x\in F),$$
$$|g_1(x)| \leqslant \tfrac{1}{3}M \quad (x\in X).$$

Making $f-g_1$ play the role of f in Lemma 2, we next find a real function g_2-g_1, continuous on X, such that

$$|f(x)-g_2(x)| \leqslant (\tfrac{2}{3})^2 M \quad (x\in F),$$
$$|g_2(x)-g_1(x)| \leqslant \tfrac{1}{2}(\tfrac{2}{3})^2 M \quad (x\in X).$$

And so on. An easy induction shows that this leads to a sequence of functions $\{g_n(x)\}$, each continuous on X, such that

$$|f(x)-g_n(x)| \leqslant (\tfrac{2}{3})^n M \quad (x\in F),$$
$$|g_n(x)-g_{n-1}(x)| \leqslant \tfrac{1}{2}(\tfrac{2}{3})^n M \quad (x\in X).$$

It follows that, if $m > n$,

$$|g_m(x)-g_n(x)| \leqslant \sum_{n+1}^{m} |g_r(x)-g_{r-1}(x)| \leqslant \tfrac{1}{2}M\sum_{n+1}^{m}(\tfrac{2}{3})^r < (\tfrac{2}{3})^n M.$$

Hence for every positive value of ϵ, there exists an integer n_0 such that $|g_m(x)-g_n(x)| < \epsilon$ whenever $m > n \geqslant n_0$.

For every $x\in X$, the sequence $\{g_n(x)\}$ is a Cauchy sequence of real numbers and so is convergent. The limit $g(x)$ therefore exists for all $x\in X$, and $g(x) = f(x)$ on F. Moreover, the Cauchy sequence is a uniformly convergent sequence of real functions, each continuous on X; hence $g(x)$ is continuous on X, which proves the theorem.

The restriction that f be bounded on F can be removed by considering for example the function $\tanh f(x)$.

EXERCISES

1. A and B are sets; f is a mapping of A into B, g is a mapping of B into A. If the mapping $g\circ f$ of A into A is a surjection and the mapping $f\circ g$ of B into B an injection, prove that f and g are both bijections.

2. The identity mapping $i:X \to X$ is defined by $i(x) = x$ for every point x of X. If $f:X \to X$ is any mapping, prove that $f \circ i = i \circ f = f$. Show also that, if $f:X \to X$ is a bijection, $f \circ f^{-1} = f^{-1} \circ f = i$ and that there is then no mapping $g:X \to X$ such that $f \circ g = g \circ f = i$ other than f^{-1}.

3. Prove that f is a continuous mapping of the metric space X into the metric space Y if and only if $\overline{f^{-1}(B)} \subseteq f^{-1}(\overline{B})$ for every subset B of Y.

4. Prove that f is a continuous mapping of the metric space X into the metric space Y if and only if $f^{-1}(\operatorname{Int} B) \subseteq \operatorname{Int} f^{-1}(B)$ for every subset B of Y.

5. $f:X \to Y$ is a mapping of a metric space X into a metric space Y, and A and B are open subsets of X. Prove that, if f is continuous on A and on B, it is continuous on $A \cup B$. Is this result true if A and B are closed subsets? Is the result true for the union of (i) an infinite number of open sets, (ii) an infinite number of closed sets?

6. $f:X \to Y$ is a mapping of a metric space X into a metric space Y. The point a is an adherent point of a subset A of X, and f is continuous at a. Prove that $f(a)$ is an adherent point of $f(A)$.

Prove that f is a continuous mapping if and only if

$$f(\overline{A}) \subseteq \overline{f(A)}$$

for every subset A of X.

7. Prove that a necessary and sufficient condition for the bijection $f:X \to Y$ to be a homeomorphism is that $f(\overline{A}) = \overline{f(A)}$ for every subset A of X.

8. f is the continuous mapping of the real line into the real line defined by $f(x) = e^{-x^2}$. Find (i) an open set O such that $f(O)$ is not open, (ii) a closed set F such that $f(F)$ is not closed, (iii) a set A such that $f(\overline{A}) \subset \overline{f(A)}$.

9. If f is a continuous mapping of the metric space X into the metric space Y and if g is a continuous mapping of Y into Z, prove that $g \circ f$ is a continuous mapping of X into Z. Show that the result also holds if 'continuous' is replaced by 'uniformly continuous'.

10. f is a real valued function defined on a metric space X and continuous on an open subset O of X. If f is positive at the point a of O, prove that there is a sphere $N(a; r)$ on which f is positive.

11. A is a non-empty subset of a metric space X, and $d(x, A)$ is the distance of the point x from A. Prove that $d(x, A)$ is uniformly continuous.

A and B are separated non-empty subsets of X. Show that

$$\{x: d(x, A) < d(x, B)\}, \quad \{x: d(x, B) < d(x, A)\}$$

are disjoint open sets containing A and B respectively, and that the set

$$\{x: d(x, A) = d(x, B)\}$$

is closed.

12. $f(x, y)$ is a continuous real function of two real variables x and y. Prove that the curve in the Euclidean plane whose equation is $f(x, y) = 0$ is a closed set.

13. Prove that there is no homeomorphism of the closed interval $-1 \leqslant t \leqslant 1$ of the real line onto the closed disc $|z| \leqslant 1$ of the complex plane.

14. M is the subset of $C[0,1]$ such that $x(0) = 0$, $x(1) = 1$, and

$$\sup |x(t)| \leqslant 1.$$

Prove that
$$f(x) = \int_0^1 \{x(t)\}^2\, dt$$

is a continuous mapping of M into $[0,1]$. By considering the sequence $\{t^n\}$ of points of M, show that M is not compact.

15. f is a real continuous function defined on a metric space X. At each point of a subset A of X, $f(x)$ is either equal to $+1$ or to -1. Show that A is connected, if and only if every such $f(x)$ is constant on A. Deduce that, if $\{A_\alpha : \alpha \in S\}$ is a family of connected sets in X such that $\bigcap_{\alpha \in S} A_\alpha$ is not empty, then $\bigcup_{\alpha \in S} A_\alpha$ is connected.

16. $\{F_n\}$ is a sequence of non-empty closed subsets of a compact metric space X such that $F_{n+1} \subseteq F_n$ for all n. Prove that $F = \bigcap_1^\infty F_n$ is closed and not empty. Let $f: X \to Y$ be a continuous mapping of X into a metric space Y. If $y \in \bigcap_1^\infty f(F_n)$, prove that there is a subsequence $\{F_{k_n}\}$ of $\{F_n\}$ and a corresponding sequence of points $\{x_{k_n}\}$ such that $x_{k_n} \in F_{k_n}$, $f(x_{k_n}) = y$, and $\{x_{k_n}\}$ converges to a point x. Prove that $x \in F_n$ for every integer n, and deduce that $\bigcap_1^\infty f(F_n) \subseteq f(F)$.

17. M is a metric space consisting of a set E with metric ρ. $M \times M$ is the set $E \times E$ with metric

$$\rho_1((x_1, y_1), (x_2, y_2)) = \max\{\rho(x_1, x_2), \rho(y_1, y_2)\}.$$

Prove that the mapping $f: M \times M \to R$ defined by $f(x, y) = \rho(x, y)$ is uniformly continuous.

18. $f: X \to X$ is a mapping of a compact metric space X with metric ρ into itself, and a_0 and b_0 are any two points of X. Sequences $\{a_n\}$ and $\{b_n\}$ are defined by $a_n = f(a_{n-1})$, $b_n = f(b_{n-1})$. Prove that there is an increasing sequence $\{k_n\}$ of positive integers such that $\{a_{k_n}\}$ and $\{b_{k_n}\}$ are convergent.

If the mapping is such that

$$\rho(a, b) \leqslant \rho(f(a), f(b)),$$

show that, for every positive value of ϵ, there exists an integer n_0 such that, whenever $n > n_0$,

$$\rho(a_{l_n}, a_0) < \epsilon, \quad \rho(b_{l_n}, b_0) < \epsilon,$$

where $l_n = k_n - k_{n_0}$. Hence prove that

$$\rho(a_1, b_1) \leqslant \rho(a_{l_n}, b_{l_n}) < \rho(a_0, b_0) + 2\epsilon.$$

Deduce that, for every pair of points of X, $\rho(a, b) = \rho(f(a), f(b))$.

The sequence $\{a_{l_n}\}$ converges to a_0. Prove that the sequence $\{a_{l_n - 1}\}$ is also a convergent sequence with limit a', and that $f(a') = a_0$.

CHAPTER 8

SOME APPLICATIONS

70. Fixed point theorems

This chapter is concerned with applications of the theory of metric spaces to algebra and analysis. All depend on fixed point theorems.

To illustrate what is meant by a fixed point property, let us consider the homographic transformation

$$w = \frac{az+b}{cz+d},$$

where a, b, c, d are real or complex numbers such that $ad-bc \neq 0$. This transformation is a bijection of the extended complex plane onto itself. A point whose position is not changed by the transformation is called a fixed point. There are two fixed points, whose affixes satisfy the quadratic equation

$$cz^2+(d-a)z-b = 0;$$

if $c = 0$, one of the fixed points is the point at infinity. The same transformation occurs in projective geometry; the fixed points of a homography on a straight line or conic are called self-corresponding points or united points.

The problem of solving an algebraic equation $f(z) = 0$ can be expressed as a fixed point problem. The relation

$$w = z+f(z)$$

maps the complex plane into itself; and the fixed points are the zeros of $f(z)$. Again, consider the system of n linear equations in n variables

$$c_r = \sum_1^n a_{r,s}x_s \quad (r = 1, 2, 3, \ldots, n)$$

or, in vector notation, $\quad A\mathbf{x} = \mathbf{c},$

where A is the square matrix with $a_{r,s}$ in its rth row and sth column, and $\mathbf{x}$ and $\mathbf{c}$ are column-vectors

$$\mathbf{x} = \begin{bmatrix} x_1 \\ x_2 \\ \cdots \\ x_n \end{bmatrix}, \quad \mathbf{c} = \begin{bmatrix} c_1 \\ c_2 \\ \cdots \\ c_n \end{bmatrix}$$

Denote by I the unit matrix with $\delta_{r,s}$ in its rth row and sth column, where $\delta_{r,s} = 0$ or 1 according as $r \neq s$ or $r = s$. Then the transformation

$$\mathbf{y} = (I - A)\mathbf{x} + \mathbf{c},$$

maps the space of vectors into itself; if we put $\mathbf{y} = \mathbf{x}$ to get the fixed points, we find that they satisfy $A\mathbf{x} = \mathbf{c}$. If A is not a singular matrix, there is one fixed point; but if A is singular, either there is no fixed point or an infinite number.

71. Contraction mappings

It is often possible to prove the existence of a fixed point under a given mapping by showing that the mapping is a contraction mapping. An example of this occurs in Cauchy's method of successive approximations for proving the existence of a unique solution of a first-order ordinary differential equation. A more elementary example, which occurs in the first chapter of Bromwich's *Introduction to the Theory of Infinite Series*, is the numerical solution of a real algebraic equation $x = f(x)$. This is the problem of finding the fixed points of the mapping $y = f(x)$ of the real line into itself. The method is to construct a sequence $\{a_n\}$ of real numbers, defined by $a_{n+1} = f(a_n)$; if the sequence converges to the limit α, then α is a root of the equation $x = f(x)$.

Let $f(x)$ be continuous in a closed interval $[a, b]$ which contains the root α, and let f map $[a, b]$ into $[a, b]$. If $f(x)$ is differentiable in (a, b), the condition $|f'(x)| \leqslant k$, where $k < 1$, ensures that the method leads to the solution. For, if x_1 and x_2 are any two points of $[a, b]$, we have

$$f(x_1) - f(x_2) = f'(\xi)(x_1 - x_2),$$

where ξ lies between x_1 and x_2, and so

$$|f(x_1) - f(x_2)| \leqslant k|x_1 - x_2|. \tag{*}$$

It follows that

$$|a_{n+1}-\alpha| = |f(a_n)-f(\alpha)| \leqslant k|a_n-\alpha|,$$

and therefore $$|a_n-\alpha| \leqslant k^{n-1}|a_1-\alpha|.$$

Since $0 < k < 1$, $a_n \to \alpha$ as $n \to \infty$.

Conversely, the condition (*) implies the existence of a root. For

$$|a_{n+1}-a_n| \leqslant k|a_n-a_{n-1}| \leqslant k^{n-1}|a_2-a_1|,$$

and, so, if $m > n$,

$$\begin{aligned}|a_m-a_n| &\leqslant |a_m-a_{m-1}|+|a_{m-1}-a_{m-2}|+\ldots+|a_{n+1}-a_n|\\ &\leqslant (k^{n-1}+k^n+\ldots+k^{m-2})|a_2-a_1|\\ &< \frac{k^{n-1}}{1-k}|a_2-a_1|.\end{aligned}$$

Since $0 < k < 1$, the sequence $\{a_n\}$ is a Cauchy sequence of real numbers and therefore converges to a limit β. Since $a_{n+1} = f(a_n)$ where $f(x)$ is continuous, β is a root of $x = f(x)$.

The relation $y = f(x)$ provided a continuous mapping of the closed interval $[a, b]$ of the real line into itself. If y_1 and y_2 are the images of the points x_1 and x_2 of the closed interval,

$$|y_1-y_2| \leqslant k|x_1-x_2|,$$

where $0 < k < 1$. The distance between y_1 and y_2 is definitely less than the distance between x_1 and x_2; this is why the mapping is called a contraction mapping.

A condition such as

$$|f(x_1)-f(x_2)| \leqslant k|x_1-x_2| \quad (k > 0),$$

is called a Lipschitz condition. It implies that $f(x)$ is continuous, but it does not imply that $f(x)$ is differentiable. Moreover not all continuous functions satisfy a Lipschitz condition.

72. A fixed point theorem for metric spaces

Let N_0 be a sphere $N(a; \delta)$ in the complete metric space X. Let f be a contraction mapping of N_0 into X which satisfies the Lipschitz condition

$$\rho(f(x_1), f(x_2)) \leqslant k\rho(x_1, x_2)$$

for every pair of points x_1 and x_2 of N_0, k being a constant such that $0 < k < 1$. Then if $\rho(a, f(a)) < \delta(1-k)$, there is a unique point α in N_0 such that $\alpha = f(\alpha)$.

Starting with the point $x_1 = f(a)$, we define a sequence $\{x_n\}$ by the relation

$$x_{n+1} = f(x_n).$$

We show that each point of the sequence belongs to N_0 and that it is a Cauchy sequence.

The point x_1 belongs to N_0, since

$$\rho(a, x_1) = \rho(a, f(a)) < \delta(1-k) < \delta.$$

Suppose that $x_1, x_2, \ldots, x_r$ are points of N_0; then $x_{r+1} = f(x_r)$ is defined. If $s \leqslant r$,

$$\begin{aligned}\rho(x_s, x_{s+1}) = \rho(f(x_{s-1}), f(x_s)) &\leqslant k\rho(x_{s-1}, x_s)\\ &\leqslant k^s\rho(a, x_1) < k^s\delta(1-k).\end{aligned}$$

Since $0 < k < 1$,

$$\begin{aligned}\rho(a, x_{r+1}) &\leqslant \rho(a, x_1) + \rho(x_1, x_2) + \ldots + \rho(x_{r-1}, x_r) + \rho(x_r, x_{r+1})\\ &< (1 + k + k^2 + \ldots + k^r)\,\delta(1-k) < \delta\end{aligned}$$

so that x_{r+1} belongs to N_0. By induction $\{x_n\}$ is a sequence of points of N_0.

Next remembering that $0 < k < 1$, we find that, if $m > n$,

$$\begin{aligned}\rho(x_n, x_m) &\leqslant \rho(x_n, x_{n+1}) + \rho(x_{n+1}, x_{n+2}) + \ldots + \rho(x_{m-1}, x_m)\\ &< (k^n + k^{n+1} + \ldots + k^{m-1})\,\delta(1-k) < \delta k^n.\end{aligned}$$

But since $k^n \to 0$ as $n \to \infty$, for every positive value of ϵ we can find a positive integer n_0 such that $\delta k^{n_0} < \epsilon$. Hence if $m > n \geqslant n_0$, $\rho(x_n, x_m) < \epsilon$. The sequence $\{x_n\}$ is thus a Cauchy sequence; since X is complete, this sequence converges to a point α of X. But

$$\begin{aligned}\rho(a, \alpha) = \lim_{n\to\infty} \rho(a, x_n) &\leqslant \lim_{n\to\infty} (1 + k + k^2 + \ldots + k^{n-1})\,\rho(a, x_1)\\ &= \rho(a, x_1)/(1-k) < \delta;\end{aligned}$$

hence α is a point of N_0.

Lastly

$$\rho(x_n, f(\alpha)) = \rho(f(x_{n-1}), f(\alpha)) \leqslant k\rho(x_{n-1}, \alpha) \to 0$$

as $n \to \infty$. Therefore $\rho(\alpha, f(\alpha)) = 0$, and so $\alpha = f(\alpha)$. This proves that α is a fixed point.

It might be thought that if we had started forming the sequence with some point other than a, we might get a different fixed point. This is not so; for if the mapping had two fixed points α and β, we should have

$$0 < \rho(\alpha,\beta) = \rho(f(\alpha),f(\beta)) \leqslant k\rho(\alpha,\beta),$$

which is impossible since $k < 1$. Hence the fixed point is unique.

Sometimes it happens that f is not a contraction mapping, yet $f \circ f$ or some composition of higher order is. Suppose that g, the composition

$$f \circ f \circ f \circ \ldots f$$

with n factors, is a contraction mapping; then

$$\rho(g(x_1), g(x_2)) \leqslant k\rho(x_1, x_2)$$

for every pair of points x_1 and x_2 of N_0, k being a constant such that $0 < k < 1$. If $\rho(a, g(a)) < \delta(1-k)$, there is a unique point α in N_0 such that $\alpha = g(\alpha)$. This point α turns out to be a fixed point of the mapping f and is unique.

For let $f(\alpha) = \beta$. Then

$$g(\beta) = g \circ f(\alpha) = f \circ g(\alpha) = f(\alpha) = \beta$$

so that β is also a fixed point of g. But g has but one fixed point α, and so $\beta = \alpha$; that is, $f(\alpha) = \alpha$. The mapping f has no other fixed point, since a fixed point of f is a fixed point of g.

73. Numerical solution of equations

We saw in §71 that the iterative process $x_{n+1} = f(x_n)$ leads to a solution of the equation $x = f(x)$ when the mapping $y = f(x)$ of the real line into itself is a contraction mapping. There are other iterative processes for finding the real roots of an algebraic or transcendental equation which can be justified by the theory of contraction mappings; one of the most famous is the Newton–Raphson iteration

$$x_{n+1} = x_n - \frac{f(x_n)}{f'(x_n)}$$

for solving the equation $f(x) = 0$.

Consider the mapping

$$y = x - \frac{f(x)}{f'(x)}$$

of the real line into itself, where $f(x)/f'(x)$ is continuous on a closed interval $[a,b]$ and differentiable on the open interval (a,b). If x_1 and x_2 are any two points of $[a,b]$ which map into y_1 and y_2,

$$y_1-y_2 = x_1-x_2-\left\{\frac{f(x_1)}{f'(x_1)}-\frac{f(x_2)}{f'(x_2)}\right\}$$

$$= (x_1-x_2)\frac{f(\xi)f''(\xi)}{\{f'(\xi)\}^2},$$

where ξ lies between x_1 and x_2. Hence if

$$\left|\frac{f(x)f''(x)}{\{f'(x)\}^2}\right| \leqslant k < 1$$

on (a,b), the mapping is a contraction mapping of $[a,b]$ onto a closed interval of the real line. It follows from the theorem that if x_0 is a point of (a,b) such that $|f(x_0)/f'(x_0)| < \delta(1-k)$, the mapping has a unique fixed point α in $(x_0-\delta, x_0+\delta) \cap [a,b]$, and that the Newton–Raphson sequence converges to α.

74. Systems of linear equations

Consider a system of n linear equations in n unknowns, which it is convenient to write in the form

$$x_r = \sum_{s=1}^{n} a_{r,s}x_s + c_r \quad (r = 1, 2, \ldots, n).$$

The constants $a_{r,s}$ and c_r and the unknowns x_r are complex numbers. This system can be written in vector form

$$\mathbf{x} = A\mathbf{x}+\mathbf{c}$$

or

$$(I-A)\mathbf{x} = \mathbf{c},$$

where $\mathbf{x} = [x_1, x_2, \ldots, x_n]$ and $\mathbf{c} = [c_1, c_2, \ldots, c_n]$ are column vectors, I is the unit matrix and A is the matrix with $a_{r,s}$ as the element in its rth row and sth column. It is well-known that the system has a unique solution if and only if the matrix $I-A$ is non-singular. It is of interest to see what the fixed point theorem tells us about the problem.

The set of all column vectors $\mathbf{x}$ can be turned into a metric space in various ways. If we write

$$\mathbf{x}^1 = [x_1^1, x_2^1, \ldots, x_n^1], \quad \mathbf{x}^2 = [x_1^2, x_2^2, \ldots, x_n^2],$$

where the superscripts are merely labels, a simple metric is

$$\rho(\mathbf{x}^1, \mathbf{x}^2) = \max\{|x_r^1 - x_r^2| : r = 1, 2, \ldots, n\},$$

and the resulting metric space is complete. We wish to show that, under certain conditions, the mapping

$$\mathbf{y} = \mathbf{f}(\mathbf{x}), \quad \text{where} \quad \mathbf{f}(\mathbf{x}) = A\mathbf{x} + \mathbf{c},$$

from this metric space into itself has a fixed point. We have

$$\begin{aligned}\rho(\mathbf{f}(\mathbf{x}^1), \mathbf{f}(\mathbf{x}^2)) &= \max\left\{\left|\sum_{s=1}^{n} a_{r,s}(x_s^1 - x_s^2)\right|\right\} \\ &\leqslant \max\left\{\sum_{s=1}^{n} |a_{r,s}| \cdot |x_s^1 - x_s^2|\right\} \\ &\leqslant \rho(\mathbf{x}^1, \mathbf{x}^2) \max\left\{\sum_{s=1}^{n} |a_{r,s}| : r = 1, 2, \ldots, n\right\}.\end{aligned}$$

Hence, if, for every value of r,

$$\sum_{s=1}^{n} |a_{r,s}| \leqslant k < 1,$$

the mapping is a contraction mapping.

Let $\mathbf{a}$ be any point of this metric space, which is not a fixed point. Then $\rho(\mathbf{a}, \mathbf{f}(\mathbf{a})) = \Delta$, where $\Delta > 0$. If we choose

$$\delta > \Delta/(1-k),$$

the fixed point theorem shows that the mapping has a unique fixed point in $N(\mathbf{a}; \delta)$. As δ can be as large as we please, the mapping has precisely one fixed point, and so the system of equations has a unique solution.

Actually this is not a good result. The condition

$$\max\left\{\sum_{s=1}^{n} |a_{r,s}|\right\} < 1$$

is known to imply that all the latent roots of A lie in $|z| < 1$. Hence the latent roots of $I - A$ lie in $|z-1| < 1$. Thus $z = 0$ is not a latent root of $I - A$, so that $I - A$ is non-singular. But there are matrices A, such that $I - A$ is non-singular, which do not satisfy the condition.

A different result is obtained if we use the metric $\rho(\mathbf{x}^1, \mathbf{x}^2)$ defined by

$$\{\rho(\mathbf{x}^1, \mathbf{x}^2)\}^2 = \sum_{r=1}^{n} |x_r^1 - x_r^2|^2.$$

The resulting metric space is again complete. We now have

$$\begin{aligned}\{\rho(\mathbf{f}(\mathbf{x}^1), \mathbf{f}(\mathbf{x}^2))\}^2 &= \sum_{r=1}^{n} \left| \sum_{s=1}^{n} a_{r,s}(x_s^1 - x_s^2) \right|^2 \\ &\leqslant \sum_{r=1}^{n} \left\{ \sum_{s=1}^{n} |a_{r,s}| \cdot |x_s^1 - x_s^2| \right\}^2 \\ &\leqslant \sum_{r=1}^{n} \sum_{s=1}^{n} |a_{r,s}|^2 \cdot \sum_{t=1}^{n} |x_t^1 - x_t^2|^2 \\ &= \{\rho(\mathbf{x}^1, \mathbf{x}^2)\}^2 \sum_{r,s} |a_{r,s}|^2.\end{aligned}$$

Hence if

$$\sum_{r,s} |a_{r,s}|^2 = k^2 < 1$$

the mapping is a contraction mapping, and the proof of the existence of a unique fixed point follows as before.

This too, is not a good result. It can be shown that, if $\lambda_1, \lambda_2, \ldots, \lambda_n$ are the latent roots of A,

$$|\lambda_1|^2 + |\lambda_2|^2 + \ldots + |\lambda_n|^2 \leqslant \sum_{r,s} |a_{r,s}|^2.$$

Hence, if $k < 1$, all the latent roots of A lie in $|z| < 1$, which implies as before that $I - A$ is non-singular; but not all non-singular matrices satisfy the condition.

75. Infinite systems of linear equations

Turn now to the infinite system of linear equations

$$x_r = \sum_{s=1}^{\infty} a_{r,s} x_s + c_r \quad (r = 1, 2, 3, \ldots)$$

which we write in vector form

$$\mathbf{x} = A\mathbf{x} + \mathbf{c}.$$

We are now working in a sequence-space, and the results depend on the particular complete metric sequence-space we choose.

Consider first the complete metric space m of bounded sequences $\mathbf{x} = \{x_n\}$ of complex numbers with metric

$$\rho(\mathbf{x}^1, \mathbf{x}^2) = \sup\{|x_r^1 - x_r^2| : r \in N\}.$$

In order that the equation may have a meaning, the infinite series

$$\sum_{s=1}^{\infty} a_{r,s} x_s$$

must converge for every positive integer r and for every point x of m. Now for every fixed value of r, we can take the bounded sequence defined by

$$x_s = \frac{|a_{r,s}|}{a_{r,s}} \quad (a_{r,s} \neq 0),$$

$$x_s = 1 \qquad (a_{r,s} = 0).$$

The series $\sum_{s=1}^{\infty} |a_{r,s}|$ must therefore be convergent. Thus, in order that $A\mathbf{x}$ be meaningful for every point $\mathbf{x}$ of m it is necessary that the series

$$\sum_{s=1}^{\infty} |a_{r,s}|$$

be convergent for every positive integer r; and the condition

$$\sup_r \sum_{s=1}^{\infty} |a_{r,s}| < \infty$$

then ensures that $A\mathbf{x}$ belongs to m.

Suppose that the infinite matrix A satisfies these conditions, and consider the mapping $\mathbf{y} = \mathbf{f}(\mathbf{x})$ of m into m, where

$$\mathbf{f}(\mathbf{x}) = A\mathbf{x} + \mathbf{c}.$$

We then have

$$\begin{aligned}\rho(\mathbf{f}(\mathbf{x}^1), \mathbf{f}(\mathbf{x}^2)) &= \sup\left\{\left|\sum_{s=1}^{\infty} a_{r,s}(x_s^1 - x_s^2)\right| : r = 1, 2, 3, \ldots\right\} \\ &\leqslant \sup\left\{\sum_{s=1}^{\infty} |a_{r,s}| \cdot |x_s^1 - x_s^2|\right\} \\ &\leqslant \rho(\mathbf{x}^1, \mathbf{x}^2) . \sup\left\{\sum_{s=1}^{\infty} |a_{r,s}|\right\}.\end{aligned}$$

Hence if there exists a positive number k, between 0 and 1, such that

$$\sum_{s=1}^{\infty} |a_{r,s}| \leqslant k$$

for every positive integer r, the mapping is a contraction mapping. The existence of a unique fixed point then follows as in the finite case. Thus if $\mathbf{c}$ belongs to the space m of bounded sequences and if there exists a positive number k $(0 < k < 1)$ such that

$$\sum_{s=1}^{\infty} |a_{r,s}| \leqslant k \quad (r = 1, 2, 3, \ldots)$$

there exists a unique bounded sequence $\mathbf{x}$ which satisfies the equation $\mathbf{x} = A\mathbf{x} + \mathbf{c}$.

Next suppose that we are seeking a solution in the complete metric space l^2, which consists of sequences $\mathbf{x} = \{x_n\}$ for which $\Sigma |x_n|^2$ is convergent, the metric being defined by

$$\{\rho(\mathbf{x}^1, \mathbf{x}^2)\}^2 = \sum_{1}^{\infty} |x_r^1 - x_r^2|^2.$$

In order that the equation may have a meaning, $\mathbf{x}$, $\mathbf{c}$ and $A\mathbf{x}$ must belong to l^2. In the first place, $A\mathbf{x}$ must exist; hence the series

$$\sum_{s=1}^{\infty} a_{rs} x_s$$

must converge for every positive integer r and for every point $\mathbf{x}$ of l^2. The series must be absolutely convergent. And it is known that $\sum_{s=1}^{\infty} |a_{r,s}|\,|x_s|$ is convergent whenever $\sum_{s=1}^{\infty} |x_s|^2$ is convergent if and only if $\sum_{s=1}^{\infty} |a_{r,s}|^2$ is convergent. Thus $A\mathbf{x}$ exists if and only if each of the series

$$\sum_{s=1}^{\infty} |a_{r,s}|^2 \quad (r = 1, 2, 3, \ldots)$$

is convergent. If $\mathbf{y} = A\mathbf{x}$, we then have

$$|y_r|^2 \leqslant \sum_{s=1}^{\infty} |a_{r,s}|^2 \sum_{t=1}^{\infty} |x_t|^2 = \{\rho(\mathbf{x}, \mathbf{0})\}^2 \sum_{s=1}^{\infty} |a_{r,s}|^2 \quad (r = 1, 2, 3, \ldots).$$

Hence, if $\sum_{r=1}^{\infty}\sum_{s=1}^{\infty}|a_{r,s}|^2$ is convergent, $\mathbf{y}$ also belongs to l^2.

Suppose now that the infinite matrix A satisfies the condition that $\sum_{r,s}|a_{r,s}|^2$ is convergent; let the sum of this double series be k^2. Consider the mapping $\mathbf{y} = \mathbf{f}(\mathbf{x})$ of l^2 into l^2, where $\mathbf{f}(\mathbf{x}) = A\mathbf{x}+\mathbf{c}$. We then have

$$\{\rho(\mathbf{f}(\mathbf{x}^1), \mathbf{f}(\mathbf{x}^2))\}^2 = \sum_{r=1}^{\infty}\left|\sum_{s=1}^{\infty} a_{r,s}(x_s^1 - x_s^2)\right|^2$$
$$\leqslant \sum_{r=1}^{\infty}\left\{\sum_{s=1}^{\infty}|a_{r,s}|\,|x_s^1 - x_s^2|\right\}^2$$
$$\leqslant \sum_{r,s}|a_{r,s}|^2\sum_{t}|x_t^1 - x_t^2|^2 = k^2\{\rho(\mathbf{x}^1, \mathbf{x}^2)\}^2$$

and so
$$\rho(\mathbf{f}(\mathbf{x}^1), \mathbf{f}(\mathbf{x}^2)) \leqslant k\rho(\mathbf{x}^1, \mathbf{x}^2).$$

Therefore, if $\sum_{r,s}|a_{r,s}|^2 < 1$, the mapping is a contraction mapping. The existence of a fixed point then follows as in the finite case.

76. Systems of ordinary differential equations of the first order

The fixed point theorem enables us to prove, under suitable conditions, the existence of a set of functions

$$\{x_r(t): r = 1, 2, 3, \ldots, n\},$$

which satisfy a system of real first-order differential equations

$$\frac{dx_r}{dt} = F_r(x_1, x_2, \ldots, x_n, t)$$

with the initial conditions $x_r = a_r$ when $t = t_0$, for $r = 1, 2, \ldots, n$. Using the notation of vectors, we write

$$\mathbf{x} \quad \text{for} \quad (x_1, x_2, \ldots, x_n),$$
$$F_r(\mathbf{x}, t) \quad \text{for} \quad F_r(x_1, x_2, \ldots, x_n, t), \quad \text{and}$$
$$\mathbf{F}(\mathbf{x}, t) \quad \text{for} \quad \{F_1(\mathbf{x}, t), F_2(\mathbf{x}, t) \ldots, F_n(\mathbf{x}, t)\}.$$

The system then becomes

$$\frac{d\mathbf{x}}{dt} = \mathbf{F}(\mathbf{x}, t),$$

with initial condition $\mathbf{x} = \mathbf{a}$. We also write

$$|\mathbf{x}|^2 = \sum_{r=1}^{n} x_r^2, \quad |\mathbf{F}(\mathbf{x}, t)|^2 = \sum_{r=1}^{n} \{F_r(\mathbf{x}, t)\}^2.$$

If we denote the set of all $\mathbf{x}$ by X we can regard $\mathbf{F}$ as a mapping whose domain and range are both in X, a mapping which depends on a parameter t; or, if we denote the set of all real numbers t by T, a mapping whose domain is in $X \times T$ and whose range is in X. When we say that $\mathbf{F}$ is continuous, we mean that each component $F_r(x_1, x_2, \ldots, x_n, t)$ of $\mathbf{F}$ is a continuous function of $n+1$ variables in the ordinary sense.

Let $\mathbf{F}(\mathbf{x}, t)$ *be continuous in the closed set* $|t-t_0| \leqslant \gamma$, $|\mathbf{x}-\mathbf{a}| \leqslant \delta$ *where it satisfies a Lipschitz condition*

$$|\mathbf{F}(\mathbf{x}^1, t) - \mathbf{F}(\mathbf{x}^2, t)| < K|\mathbf{x}^1 - \mathbf{x}^2|$$

K being a positive constant. Then, if γ is sufficiently small, the differential equation

$$\frac{d\mathbf{x}}{dt} = \mathbf{F}(\mathbf{x}, t)$$

has in $|t-t_0| < \gamma$ *a unique solution which satisfies the condition* $\mathbf{x} = \mathbf{a}$ *when* $t = t_0$.

We may suppose that t_0 and all the numbers a_r are zero. Since $|\mathbf{F}|$ is continuous in $|t| \leqslant \gamma$, $|\mathbf{x}| \leqslant \delta$, it is bounded; let M be the supremum of $|\mathbf{F}|$. We may also suppose that $\gamma(M+K\delta) < \delta$; for if not, we may replace γ by a smaller number without changing δ, K or M, though M will then possibly be an upper bound and not the supremum. If we write $K\gamma = k$, we have $M\gamma < \delta(1-k)$ so that $0 < k < 1$.

If the differential equation has a solution $\mathbf{x}(t)$ which satisfies the initial condition, each $x_r(t)$ is differentiable and therefore continuous. But since

$$\frac{dx_r}{dt} = F_r(\mathbf{x}(t), t),$$

where F_r is continuous, each dx_r/dt is continuous. Therefore

$$x_r(t) = \int_0^t x_r'(\tau)\, d\tau = \int_0^t F_r(\mathbf{x}(\tau), \tau)\, d\tau.$$

Hence if the differential equation has a solution $\mathbf{x}(t)$ which vanishes with t, $\mathbf{x}(t)$ satisfies the integral equation

$$\mathbf{x}(t) = \int_0^t \mathbf{F}(\mathbf{x}(\tau), \tau)\, d\tau.$$

Conversely, if $\mathbf{x}(t)$ is a continuous solution of this integral equation, $d\mathbf{x}(t)/dt$ exists and is equal to $\mathbf{F}(\mathbf{x}(t), t)$. Thus solving the differential equation is equivalent to solving the integral equation.

Let C be the space of all $\mathbf{x}(t)$ continuous on $-\gamma \leqslant t \leqslant \gamma$, with metric

$$\rho(\mathbf{x}^1, \mathbf{x}^2) = \sup\{|\mathbf{x}^1(t) - \mathbf{x}^2(t)| : -\gamma \leqslant t \leqslant \gamma\}.$$

This space is complete; for if $\{\mathbf{x}^p(t)\}$ is a Cauchy sequence with respect to this metric, each $x_r^p(t)$ is uniformly convergent in the ordinary sense as $p \to \infty$. Let N_0 be the sphere $N(\mathbf{0}; \delta)$ where $\mathbf{0} = (0, 0, 0, \ldots, 0)$; this is the set of all points $\mathbf{x}(t)$ of C such that $\sup |\mathbf{x}(t)| < \delta$. We have to show that the mapping $\mathbf{y} = \mathbf{f}(\mathbf{x})$ of N_0 into X defined by

$$\mathbf{y}(t) = \int_0^t \mathbf{F}(\mathbf{x}(\tau), \tau)\, d\tau$$

has a fixed point.

First, this is a mapping of N_0 into N_0. For if $0 \leqslant t \leqslant \gamma$,

$$\begin{aligned} |\mathbf{y}(t)|^2 &= \sum_{r=1}^{n} \left\{\int_0^t F_r(\mathbf{x}(\tau), \tau)\, d\tau\right\}^2 \\ &\leqslant \sum_{r=1}^{n} \left\{\int_0^t |F_r(\mathbf{x}(\tau), \tau)|\, d\tau\right\}^2 \\ &\leqslant \sum_{r=1}^{n} t \int_0^t |F_r(\mathbf{x}(\tau), \tau)|^2\, d\tau \\ &\leqslant M^2 t^2 \leqslant M^2 \gamma^2; \end{aligned}$$

and similarly with $-\gamma \leqslant t \leqslant 0$. Hence $|\mathbf{y}(t)| \leqslant M\gamma < \delta(1-k)$. Therefore

$$\sup |\mathbf{y}(t)| \leqslant \delta(1-k) < \delta,$$

as required. Incidentally we have proved that $\rho(\mathbf{0}, \mathbf{f}(\mathbf{0})) < \delta(1-k)$ which is one of the conditions of the fixed point theorem.

The mapping is also a contraction mapping. For if $\mathbf{y}^1(t)$ and

$\mathbf{y}^2(t)$ are the images of the points $\mathbf{x}^1(t)$ and $\mathbf{x}^2(t)$ of N_0, and if $0 \leqslant t \leqslant \gamma$, we have

$$\begin{aligned}|\mathbf{y}^1(t)-\mathbf{y}^2(t)|^2 &= \sum_{r=1}^{n}\left\{\int_0^t [F_r(\mathbf{x}^1(\tau),\tau)-F_r(\mathbf{x}^2(\tau),\tau)]\,d\tau\right\}^2 \\ &\leqslant \sum_{r=1}^{n} t\int_0^t |F_r(\mathbf{x}^1(\tau),\tau)-F_r(\mathbf{x}^2(\tau),\tau)|^2\,d\tau \\ &< K^2t\int_0^t |\mathbf{x}^1(\tau)-\mathbf{x}^2(\tau)|^2\,d\tau \\ &< K^2t^2\sup|\mathbf{x}^1(\tau)-\mathbf{x}^2(\tau)|^2 \\ &\leqslant K^2\gamma^2\{\rho(\mathbf{x}^1,\mathbf{x}^2)\}^2 = k^2\{\rho(\mathbf{x}^1,\mathbf{x}^2)\}^2\end{aligned}$$

and similary if $-\gamma \leqslant t \leqslant 0$. Hence

$$\rho(\mathbf{y}^1,\mathbf{y}^2) \leqslant k\rho(\mathbf{x}^1,\mathbf{x}^2),$$

where $0 < k < 1$.

The conditions of the fixed point theorem are thus satisfied. The mapping has a unique fixed point $\boldsymbol{\alpha}(t)$, and so

$$\boldsymbol{\alpha}(t) = \int_0^t \mathbf{F}(\boldsymbol{\alpha}(\tau),\tau)\,d\tau.$$

The system of differential equations

$$\frac{d\mathbf{x}}{dt} = \mathbf{F}(\mathbf{x},t)$$

has therefore the unique solution $\mathbf{x} = \boldsymbol{\alpha}(t)$ on $-\gamma \leqslant t \leqslant \gamma$, which satisfies the initial condition $\mathbf{x} = \mathbf{0}$ when $t = 0$.

It should be noted that the differential equation of order n

$$\frac{d^nx}{dt^n} = F\left(x, \frac{dx}{dt}, \frac{d^2x}{dt^2}, \ldots, \frac{d^{n-1}x}{dt^{n-1}}, t\right)$$

can be expressed as a system of first-order equations by writing

$$x = x_1, \quad \frac{dx}{dt} = x_2, \ldots, \frac{d^{n-1}x}{dt^{n-1}} = x_n.$$

The system is

$$\frac{dx_r}{dt} = x_{r+1} \quad (r = 1, 2, \ldots, n-1),$$

$$\frac{dx_n}{dt} = F(x, x_2, \ldots, x_n, t).$$

The theorem of this section therefore includes an existence theorem for a differential equation of any order.

77. Boundary value problems

Let $F(x,y,t)$ be a real continuous function. If (x_1,y_1,t), (x_2,y_2,t) are any two points of $-\delta \leqslant x \leqslant \delta$, $-\delta \leqslant y \leqslant \delta$, $0 \leqslant t \leqslant 1$, let F satisfy a condition

$$|F(x_1,y_1,t)-F(x_2,y_2,t)| \leqslant \mu_0|x_1-x_2|+\mu_1|y_1-y_2|,$$

where μ_0, μ_1 depend only on δ. Then, for all sufficiently small values of the real constant λ, the differential equation

$$\frac{d^2x}{dt^2} = \lambda F\left(x, \frac{dx}{dt}, t\right)$$

has a unique solution, which is continuously differentiable in $0 \leqslant t \leqslant 1$, where it satisfies the conditions $|x(t)| < \delta$, $|x'(t)| < \delta$, and which vanishes where $t = 0$ and $t = 1$.

When we speak of differentiability in the closed interval $[0,1]$, we understand that the derivatives are one-sided at the end-points. Since F is continuous on $|x| \leqslant \delta$, $|y| \leqslant \delta$, $0 \leqslant t \leqslant 1$, it is bounded; let M be its supremum. Evidently M will depend on δ. The case $\lambda = 0$ is trivial.

If this boundary value problem has a continuously differentiable solution $x(t)$, the function $F(x(t), x'(t), t)$ is continuous on $[0,1]$. Hence we may integrate the expression on each side of the differential equation and obtain

$$x'(t) = A+\lambda\int_0^t F(x(v),x'(v),v)\,dv,$$

$$x(t) = At+B+\lambda\int_0^t (t-v)\,F(x(v),x'(v),v)\,dv,$$

where A and B are constants of integration. But since $x(t)$ vanishes when $t = 0$, and $t = 1$, we have

$$A+\lambda\int_0^1 (1-v)\,F(x(v),x'(v),v)\,dv = 0, \quad B = 0.$$

It follows that

$$x(t) = -\lambda(1-t)\int_0^t vF(x(v), x'(v), v)\,dv$$
$$-\lambda t\int_t^1 (1-v)\,F(x(v), x'(v), v)\,dv.$$

Conversely, if $x(t)$ is a continuously differentiable solution of this integral equation, it is a solution of the boundary value problem.

Now consider the metric space A, which consists of all real functions which vanish at $t = 0$ and are continuously differentiable on $[0, 1]$, with metric

$$\rho(x_1, x_2) = \sup\{|x_1'(t) - x_2'(t)| : t \in [0, 1]\}.$$

A is a subspace of the complete metric space $C^1[0, 1]$, and is itself complete. For if $\{x_n(t)\}$ is a Cauchy sequence, the sequence $\{x_n'(t)\}$ of continuous functions converges uniformly in the ordinary sense to a continuous function, and this implies that $\{x_n(t)\}$ converges uniformly in the ordinary sense.

Now consider the mapping f of A into itself defined by

$$y(t) = -\lambda(1-t)\int_0^t vF(x(v), x'(v), v)\,dv$$
$$-\lambda t\int_t^1 (1-v)\,F(x(v), x'(v), v)\,dv.$$

We have to show that it has a fixed point for all sufficiently small values of λ. Let $x(t)$ be a point of $N_0 = N(0; \delta)$ so that

$$|x(t)| < \delta, |x'(t)| < \delta \quad \text{when} \quad 0 \leqslant t \leqslant 1.$$

Since

$$y'(t) = \lambda\int_0^t vF(x(v), x'(v), v)\,dv - \lambda\int_t^1 (1-v)\,F(x(v), x'(v), v)\,dv,$$

we have

$$|y'(t)| \leqslant |\lambda| M\int_0^t v\,dv + |\lambda| M\int_0^1 (1-v)\,dv = \tfrac{1}{2}|\lambda| M(1 - 2t + 2t^2)$$

and so

$$\rho(0, y) = \sup\{|y'(t)| : t \in [0, 1]\} \leqslant \tfrac{1}{2}|\lambda| M < \delta(1-k),$$

where $0 < k < 1$ provided $|\lambda|$ is small enough. Then f maps N_0 into N_0.

Next, let $y_1 = f(x_1), y_2 = f(x_2)$, where x_1 and x_2 are any two points of N_0. We have

$$y_1'(t) - y_2'(t) = \lambda \int_0^t v\,G\,dv - \lambda \int_t^1 (1-v)\,G\,dv,$$

where $$G = F(x_1(v), x_1'(v), v) - F(x_2(v), x_2'(v), v).$$

Therefore $$|G(v)| \leqslant \mu_0|x_1(v) - x_2(v)| + \mu_1|x_1'(v) - x_2'(v)|.$$

But
$$\begin{aligned} |x_1(v) - x_2(v)| &= \left|\int_0^v \{x_1'(t) - x_2'(t)\}\,dt\right| \\ &\leqslant \sup|x_1'(t) - x_2'(t)| \int_0^v dt \\ &\leqslant \sup|x_1'(t) - x_2'(t)|. \end{aligned}$$

Therefore $$|G(v)| \leqslant (\mu_0 + \mu_1)\rho(x_1, x_2).$$

Hence

$$\begin{aligned} |y_1'(t) - y_2'(t)| &\leqslant (\mu_0 + \mu_1)\rho(x_1, x_2)\,|\lambda| \left\{\int_0^t v\,dv + \int_t^1 (1-v)\,dv\right\} \\ &= \tfrac{1}{2}(\mu_0 + \mu_1)\rho(x_1, x_2)\,|\lambda|\,(1 - 2t + 2t^2) \\ &\leqslant \tfrac{1}{2}|\lambda|(\mu_0 + \mu_1)\rho(x_1, x_2). \end{aligned}$$

It follows that

$$\rho(f(x_1), f(x_2)) \leqslant \tfrac{1}{2}|\lambda|\,(\mu_0 + \mu_1)\rho(x_1, x_2) \leqslant k\rho(x_1, x_2)$$

if $|\lambda| \leqslant 2k/(\mu_0 + \mu_1)$.

The conditions of the fixed point theorem are thus satisfied for all sufficiently small values of $|\lambda|$. The integral equation and the associated boundary value problem then have a unique continuously differentiable solution.

If the boundary conditions were $x(0) = c_0$, $x(1) = c_1$, the proof could be carried through with slight modifications, or we could reduce the problem to the one solved here by replacing x by $x - c_0(1-t) - c_1 t$.

78. Fredholm's integral equation

Let $\phi(s)$ be continuous on the interval $a \leqslant s \leqslant b$. Let $K(s,t)$ be continuous on the square $a \leqslant s \leqslant b$, $a \leqslant t \leqslant b$. Then for all sufficiently small values of $|\lambda|$, the integral equation

$$x(s) = \phi(s) + \lambda \int_a^b K(s,t)\, x(t)\, dt$$

has a unique solution, continuous on $[a,b]$.

Let X be the complete metric space $C[a,b]$ of all continuous functions $x(s)$ with metric

$$\rho(x_1, x_2) = \sup\{|x_1(s) - x_2(s)| : a \leqslant s \leqslant b\}.$$

Since the expression on the right-hand side of the integral equation is continuous on $[a,b]$,

$$y(s) = \phi(s) + \lambda \int_a^b K(s,t)\, x(t)\, dt$$

is also a point of X. If we write $y = f(x)$, f is a mapping of X into itself. We have to prove that, for all sufficiently small values of $|\lambda|$, the mapping has a unique fixed point.

In the general theorem, take N_0 to be the sphere $N(0;\delta)$, where δ can be chosen as large as we please. Since ϕ and K are continuous, they are bounded; let $|\phi(s)| \leqslant m$, $|K(s,t)| \leqslant M$. Then, if $y_1 = f(x_1)$, $y_2 = f(x_2)$,

$$|y_1(s) - y_2(s)| = \left|\lambda \int_a^b K(s,t)\{x_1(t) - x_2(t)\}\, dt\right| \leqslant |\lambda|\, M(b-a)\, \rho(x_1, x_2),$$

and so

$$\rho(f(x_1), f(x_2)) = \sup |y_1(s) - y_2(s)| \leqslant |\lambda|\, M(b-a)\, \rho(x_1, x_2).$$

Therefore if we choose any positive number k $(0 < k < 1)$ and consider values of $|\lambda|$ such that $|\lambda|\, M(b-a) \leqslant k$, the mapping is a contraction mapping.

Again, since $f(0) = \phi(s)$,

$$\rho(0, f(0)) = \sup |\phi(s)| \leqslant m.$$

If we suppose δ chosen so that $\delta > m/(1-k)$, the second condition of the theorem is satisfied.

It follows that there is a unique point α in N_0 such that $\alpha = f(\alpha)$.

As δ can be as large as we please, this means that, for every sufficiently small value of $|\lambda|$, the integral equation has a unique continuous solution $x(s) = \alpha(s)$.

In particular, for all sufficiently small values of $|\lambda|$, the only solution of the homogeneous equation

$$x(s) = \lambda \int_a^b K(s,t)\, x(t)\, dt$$

is $x(s) \equiv 0$. For if this were not so, the solution of the non-homogeneous equation would not be unique; we could add to it any solution of the homogeneous equation.

79. Volterra's integral equation

Let $\phi(s)$ be continuous on the interval $a \leqslant s \leqslant b$. Let $K(s,t)$ be continuous on the triangle $a \leqslant t \leqslant s \leqslant b$. Then for all values of λ, the equation

$$x(s) = \phi(s) + \lambda \int_a^s K(s,t)\, x(t)\, dt \quad (a \leqslant s \leqslant b)$$

has a unique continuous solution.

This differs from the Fredholm equation in that the upper limit of the integral is the variable s and that there is no restriction on the value of λ.

Let X be the complete metric space $C[a,b]$ of all continuous functions $x(s)$ with metric

$$\rho(x_1, x_2) = \sup\{|x_1(s) - x_2(s)| : a \leqslant s \leqslant b\}.$$

Since the expression on the right-hand side of the integral equation is continuous on $[a,b]$,

$$y(s) = \phi(s) + \lambda \int_a^s K(s,t)\, x(t)\, dt$$

is also a point of X. If we write $y = f(x)$, f is a mapping of X into itself.

In the fixed point theorem, take N_0 to be the sphere $N(0;\delta)$ where δ can be chosen as large as we please. Since ϕ and K are continuous, they are bounded; let $|\phi(s)| \leqslant m$, $|K(s,t)| \leqslant M$. Then

$$|y_1(s) - y_2(s)| = \left|\lambda \int_a^s K(s,t)\{x_1(t) - x_2(t)\}\, dt\right|$$
$$\leqslant |\lambda| M(b-a)\rho(x_1, x_2)$$

and so $$\rho(f(x_1), f(x_2)) \leqslant |\lambda| M(b-a)\rho(x_1, x_2)$$

just as in the Fredholm case. Thus f is a contraction mapping only for sufficiently small values of $|\lambda|$.

Denote by $f^{(n)}$ the composition

$$f \circ f \circ \ldots \circ f$$

with n factors, and let $f^{(n)}(x) = y^{(n)}$. It is readily proved by induction that

$$y^{(n)}(s) = \phi(s) + \sum_{r=1}^{n-1} \lambda^r \int_a^s K_r(s,t)\,\phi(t)\,dt + \lambda^n \int_a^s K_n(s,t)\,x(t)\,dt,$$

where

$$K_1(s,t) = K(s,t), \quad K_{r+1}(s,t) = \int_t^s K_1(s,u)\,K_r(u,t)\,du$$

and $$a \leqslant t \leqslant s \leqslant b.$$

Now $|K_1(s,t)| \leqslant M$. Hence

$$|K_2(s,t)| = \left|\int_t^s K_1(s,u)\,K_1(u,t)\,du\right| \leqslant M^2(s-t),$$

$$|K_3(s,t)| = \left|\int_t^s K_1(s,u)\,K_2(u,t)\,du\right| \leqslant M^3 \int_t^s (u-t)\,du$$

$$= \frac{M^3(s-t)^2}{2!}$$

and, generally,

$$|K_r(s,t)| \leqslant \frac{M^r(s-t)^{r-1}}{(r-1)!}.$$

Using this bound for $|K_r|$, we have

$$|y_1^{(n)}(s) - y_2^{(n)}(s)| = \left|\lambda^n \int_a^s K_n(s,t)\,\{x_1(t) - x_2(t)\}\,dt\right|$$

$$\leqslant |\lambda^n| \int_a^s M^n \frac{(s-t)^{n-1}}{(n-1)!} |x_1(t) - x_2(t)|\,dt$$

$$\leqslant \frac{|\lambda|^n M^n}{(n-1)!} \rho(x_1, x_2) \int_a^s (s-t)^{n-1}\,dt$$

$$= \frac{1}{n!} |\lambda|^n M^n (s-a)^n \rho(x_1, x_2).$$

Therefore $\rho(y_1^{(n)}, y_2^{(n)}) = \sup\{|y_1^{(n)} - y_2^{(n)}| : a \leqslant s \leqslant b\}$

$$\leqslant \frac{1}{n!}|\lambda|^n M^n (b-a)^n \rho(x_1, x_2).$$

But $|\lambda|^n M^n(b-a)^n/n!$ tends to zero as $n \to \infty$. Hence we can choose n so that

$$k = \frac{|\lambda|^n M^n (b-a)^n}{n!}$$

lies between 0 and 1. Therefore, with such a choice of n,

$$\rho(f^{(n)}(x_1), f^{(n)}(x_2)) \leqslant k\rho(x_1, x_2),$$

where $0 < k < 1$. The mapping $f^{(n)}$ is thus a contraction mapping, for any choice of λ.

Lastly,

$$\rho(0, f^{(n)}(0)) = \sup\left|\phi(s) + \sum_{r=1}^{n-1} \lambda^r \int_a^s K_r(s,t)\,\phi(t)\,dt\right|$$

$$\leqslant m\left\{1 + \sum_{r=1}^{n-1} |\lambda|^r M^r \frac{(b-a)^r}{r!}\right\},$$

which is finite. We can therefore choose δ so large that

$$\rho(0, f^{(n)}(0)) < \delta(1-k).$$

The conditions of the fixed point theorem are satisfied. The mapping $y = f^{(n)}(x)$ has therefore a unique fixed point $\alpha(s)$ in N_0. As we saw in §72 it follows that $\alpha(s)$ is a unique fixed point of the mapping $y = f(x)$. Hence Volterra's integral equation has a unique continuous solution, irrespective of the value of λ.

80. The fixed point theorem with a parameter

Let X be a complete metric space with metric ρ_1; let N_1 be the sphere $N_1(a;\delta)$. Let T be a metric space with metric ρ_2; let N_2 be the sphere $N_2(b;\gamma)$. Let f be a continuous mapping of $N_1 \times N_2$ into X which satisfies the Lipschitz condition

$$\rho_1(f(x_1,t), f(x_2,t)) \leqslant k\rho_1(x_1, x_2)$$

for every pair of points (x_1, t) and (x_2, t) of $N_1 \times N_2$, k being a constant *such that* $0 < k < 1$. *If, for every point* t *of* $N_2, \rho_1(a, f(a,t)) < \delta(1-k)$, *there exists a unique mapping* $\alpha(t)$ *of* N_2 *into* N_1 *such that*

$$\alpha(t) = f(\alpha(t), t),$$

and $\alpha(t)$ *is continuous on* N_2.

The metric of $X \times T$ is any one of the equivalent metrics of §61.

Apart from the reference to the parameter t, this theorem is identical with that of §72, so that there is no need to prove the existence of a unique $\alpha(t)$ for each point t of N_2. It only remains to prove that $\alpha(t)$ is continuous.

As in §72, we form a sequence $\{x_n\}$ now defined by relations

$$x_1 = f(a,t), \quad x_{n+1} = f(x_n, t),$$

where each x_n depends on the parameter t; and the proof shows that $\{x_n(t)\}$ is uniformly convergent to $\alpha(t)$ on the closed sphere $K_2(b; \gamma_1)$, where γ_1 is any number less than γ. Since $f(x,t)$ is continuous, x_1 is continuous; it follows by induction that each member of the sequence is continuous on $K_2(b; \gamma_1)$. Therefore $\alpha(t)$, being the uniform limit of a sequence of continuous functions, is continuous on $K_2(b; \gamma_1)$ for every $\gamma_1 < \gamma$, and hence on N_2.

81. The inverse function theorem

A simple consequence of the fixed point theorem with a parameter is the inverse function theorem of elementary analysis. One form of this theorem is as follows.

Let $F(x)$ *be a real function of the real variable* x *which satisfies the condition*

$$0 < m \leqslant \frac{F(x_1) - F(x_2)}{x_1 - x_2} \leqslant M \quad (m < M)$$

for every pair of points of the interval $|x-a| < \delta$. *Let* $F(a) = b$. *Then if* $|t-b| < m\delta$, *the equation* $F(x) = t$ *has a unique solution* $x = \alpha(t)$, *where* $a = \alpha(b)$, *and* $\alpha(t)$ *is continuous on* $|t-b| < m\delta$.

We apply the fixed point theorem to

$$f(x,t) = x + \lambda\{t - F(x)\},$$

where λ is a real constant to be suitably determined, the metric spaces X and T both being the real line with the usual metric. The condition satisfied by $F(x)$ implies that $F(x)$ is continuous. Hence f is a continuous mapping of $X \times T$ into the real line.

The condition
$$\rho(f(x_1,t),f(x_2,t)) \leqslant k\rho_1(x_1,x_2)$$
becomes
$$|x_1 - x_2 - \lambda\{F(x_1) - F(x_2)\}| \leqslant k|x_1 - x_2|$$
or
$$\left|1 - \lambda\frac{F(x_1)-F(x_2)}{x_1-x_2}\right| \leqslant k$$
or, again,
$$1-k \leqslant \lambda\frac{F(x_1)-F(x_2)}{x_1-x_2} \leqslant 1+k.$$

This inequality is satisfied if we take
$$\frac{1-k}{\lambda} = m, \quad \frac{1+k}{\lambda} = M,$$
which gives
$$\lambda = \frac{2}{M+m}, \quad k = \frac{M-m}{M+m},$$
so that $0<k<1$. The other condition
$$\rho_1(a,f(a,t)) < \delta(1-k)$$
becomes
$$|t-b| < m\delta,$$
which is one of the assumptions.

The conditions of the fixed point theorem are thus satisfied. It follows that, if $|t-b| < m\delta$, where $b = F(a)$, the equation $F(x) = t$ has a unique solution $x = \alpha(t)$, where $\alpha(t)$ is continuous.

82. The implicit function theorem

This theorem is concerned with the circumstances under which a set of n equations in n unknowns $x_1, x_2, \ldots, x_n$ of the form
$$F_r(x_1, x_2. \ldots, x_n;\ t_1, t_2, \ldots, t_m) = 0 \quad (r = 1, 2, \ldots, n)$$
involving m parameters $t_1, t_2, \ldots, t_m$, can be solved. It is convenient to use vector notation. Let us regard $(x_1, x_2, \ldots, x_n)$ as the co-ordinates of a point $\mathbf{x}$ in an n-dimensional Euclidean space X and $(t_1, t_2, \ldots, t_m)$ as the co-ordinates of a point $\mathbf{t}$ in an m-dimensional

Euclidean space T. The product space $X \times T$ is an $(m+n)$ dimensional Euclidean space. We write the set of equations as

$$F_r(\mathbf{x};\ \mathbf{t}) = 0 \quad (r = 1, 2, \ldots, n)$$

or

$$\mathbf{F}(\mathbf{x};\ \mathbf{t}) = \mathbf{0},$$

where $\mathbf{F}$ is a point of X, depending on $\mathbf{x}$ and on $\mathbf{t}$.

Suppose that each component F_r of $\mathbf{F}$ is continuously differentiable on an open set S of $X \times T$. Let

$$J = \frac{\partial(F_1, F_2, \ldots, F_n)}{\partial(x_1, x_2, \ldots, x_n)}$$

be the Jacobian of $\mathbf{F}$ with respect to $\mathbf{x}$. *If* $J \neq 0$ *at a point* $(\mathbf{x}^0;\mathbf{t}^0)$ *of* S *and if* $\mathbf{F}(\mathbf{x}^0;\mathbf{t}^0) = \mathbf{0}$, *the equation* $\mathbf{F}(\mathbf{x};\ \mathbf{t}) = \mathbf{0}$ *has a unique solution* $\mathbf{x} = \boldsymbol{\alpha}(t)$ *which is continuously differentiable on a sphere* $|\mathbf{t}-\mathbf{t}^0| < \gamma$.

We may suppose $\mathbf{x}^0 = \mathbf{0}$, $\mathbf{t}^0 = \mathbf{0}$. Let $A(\mathbf{x};\ \mathbf{t})$ be the Jacobian matrix which has $\partial F_r/\partial x_s$ in its rth row and sth column; then $J = \det A$. By hypothesis, the matrix, $A_0 = A(\mathbf{0};\mathbf{0})$ is non-singular and so has an inverse A_0^{-1}. If A_0 is not the unit matrix, consider the vector

$$\mathbf{G} = A_0^{-1}\mathbf{F}.$$

The Jacobian matrix of $\mathbf{G}$ is $A_0^{-1}A$ which reduces to the unit matrix when $\mathbf{x} = \mathbf{0}$, $\mathbf{t} = \mathbf{0}$. Since $\mathbf{G} = \mathbf{0}$ if and only if $\mathbf{F} = \mathbf{0}$, it suffices to consider a system of equations $\mathbf{F} = \mathbf{0}$ whose Jacobian matrix is the unit matrix when $\mathbf{x} = \mathbf{0}, \mathbf{t} = \mathbf{0}$. Let us apply the fixed point theorem with a parameter $\mathbf{t}$ to the mapping $\mathbf{f}: S \to X$, where

$$\mathbf{f}(\mathbf{x};\mathbf{t}) = \mathbf{x} - \mathbf{F}(\mathbf{x};\mathbf{t}).$$

A fixed point of this mapping is a solution of the equation $\mathbf{F}(\mathbf{x};\ \mathbf{t}) = \mathbf{0}$.

The first component of the vector $\mathbf{f}(\mathbf{x}';\ \mathbf{t}) - \mathbf{f}(\mathbf{x};\ \mathbf{t})$ is

$$\begin{aligned} f_1(\mathbf{x}';\ \mathbf{t}) - f_1(\mathbf{x};\ \mathbf{t}) &= x_1' - x_1 - \{F_1(\mathbf{x}';\ \mathbf{t}) - F_1(\mathbf{x};\ \mathbf{t})\} \\ &= x_1' - x_1 - \sum_{r=1}^{n} \{F_1(\mathbf{x}^r;\ \mathbf{t}) - F_1(\mathbf{x}^{r+1}, \mathbf{t})\}, \end{aligned}$$

where

$$\mathbf{x}^1 = \mathbf{x}', \quad \mathbf{x}^{n+1} = \mathbf{x}$$

but

$$\mathbf{x}^r = (x_1, x_2, \ldots, x_{r-1}, x_r', x_{r+1}', \ldots, x_n')$$

if $1 < r < n+1$. Since $\mathbf{F}_1$ is differentiable,

$$F_1(\mathbf{x}^r;\, \mathbf{t}) - F_1(\mathbf{x}^{r+1};\, \mathbf{t}) = (x_r' - x_r)\, F_{1,r}(\boldsymbol{\xi}^r;\, \mathbf{t}),$$

where
$$\boldsymbol{\xi}^r = (x_1, x_2, \ldots, x_{r-1}, x_r'', x_{r+1}', \ldots, x_n');$$

here x_r'' lies between x_r and x_r' and $F_{1,r}$ $(\mathbf{x};\, \mathbf{t})$ denotes $\partial F_1/\partial x_r$. Hence

$$f_1(\mathbf{x}';\, \mathbf{t}) - f_1(\mathbf{x};\, \mathbf{t}) = \sum_{r=1}^{n} c_r(x_r' - x_r),$$

where
$$c_1 = 1 - F_{1,1}(\boldsymbol{\xi}^1;\, \mathbf{t}), \quad c_r = -F_{1,r}(\boldsymbol{\xi}^r;\, \mathbf{t}) \quad (r > 1).$$

Now $\{F_{1,r}(\mathbf{x};\, \mathbf{t}) : r = 1, 2, \ldots, n\}$ is the first row of the Jacobian matrix $A(\mathbf{x};\, \mathbf{t})$. By hypothesis, each element of A is continuous, and A tends to the unit matrix as $\mathbf{x}$ and $\mathbf{t}$ tend to zero. Therefore each of the coefficients $c_1, c_2, \ldots, c_n$ tends to zero as $\mathbf{x}, \mathbf{x}'$ and $\mathbf{t}$ tend to zero. For every positive value of ϵ, there exist positive numbers γ_1 and δ_1 such that $|c_r| < \epsilon$ whenever $|\mathbf{x}| < \delta_1$, $|\mathbf{x}'| < \delta_1$, $|\mathbf{t}| < \gamma_1$. Therefore

$$|f_1(\mathbf{x}';\, \mathbf{t}) - f_1(\mathbf{x};\, \mathbf{t})| \leqslant \left\{\sum_1^n |c_r|^2 . \sum_1^n |x_s' - x_s|^2\right\}^{\frac{1}{2}} < \epsilon|\mathbf{x}' - \mathbf{x}|\sqrt{n},$$

whenever $|\mathbf{x}| < \delta_1$, $|\mathbf{x}'| < \delta_1$, $|\mathbf{t}| < \gamma_1$.

Similarly, for every value of r, there exist positive numbers γ_r and δ_r such that

$$|f_r(\mathbf{x}';\, \mathbf{t}) - f_r(\mathbf{x};\, \mathbf{t})| < \epsilon|\mathbf{x}' - \mathbf{x}|\sqrt{n},$$

whenever $|\mathbf{x}| < \delta_r$, $|\mathbf{x}'| < \delta_r$, $|\mathbf{t}| < \gamma_r$. Hence

$$|\mathbf{f}(\mathbf{x}';\mathbf{t}) - \mathbf{f}(\mathbf{x};\mathbf{t})| < n\epsilon\,|\mathbf{x}' - \mathbf{x}|,$$

whenever $|\mathbf{x}| < \delta = \min \delta_r$, $|\mathbf{x}'| < \delta$, $|\mathbf{t}| < \gamma \leqslant \min \gamma_r$. If we choose ϵ so that $k = n\epsilon < 1$, the mapping $\mathbf{t}$ is a contraction mapping.

Next
$$\rho(\mathbf{0}, \mathbf{f}(\mathbf{0};\mathbf{t})) = |\mathbf{F}(\mathbf{0};\mathbf{t})| \to 0$$

as $|\mathbf{t}| \to 0$. We may therefore suppose that γ was chosen so small that

$$\rho(\mathbf{0}, \mathbf{f}(\mathbf{0};\mathbf{t})) < \delta(1-k),$$

whenever $|\mathbf{t}| < \gamma$, the second condition of the fixed point theorem.

It follows that the equation $\mathbf{x} = \mathbf{f}(\mathbf{x}; \mathbf{t})$, that is, $\mathbf{F}(\mathbf{x}; \mathbf{t}) = \mathbf{0}$, has a unique solution $\mathbf{x} = \boldsymbol{\alpha}(\mathbf{t})$ which is continuous in $|\mathbf{t}| < \gamma$ and satisfies the condition $\boldsymbol{\alpha}(\mathbf{0}) = \mathbf{0}$.

It remains to prove that $\boldsymbol{\alpha}(\mathbf{t})$ is continuously differentiable. Let $\mathbf{t}$ and $\mathbf{t}+\Delta\mathbf{t}$ be two points such that $|\mathbf{t}| < \gamma$, $|\mathbf{t}+\Delta\mathbf{t}| < \gamma$. Write $\mathbf{x} = \boldsymbol{\alpha}(\mathbf{t})$, $\mathbf{x}+\Delta\mathbf{x} = \boldsymbol{\alpha}(\mathbf{t}+\Delta\mathbf{t})$. Then $\mathbf{F}(\mathbf{x}+\tau\Delta\mathbf{x}; \mathbf{t}+\tau\Delta\mathbf{t})$ vanishes when $\tau = 0$ and when $\tau = 1$. Consider each component of this vector separately.

By the mean value theorem, there exists a number

$$\tau_r \quad (0 < \tau_r < 1)$$

such that
$$\frac{d}{d\tau} F_r(\mathbf{x}+\tau\Delta\mathbf{x}; \mathbf{t}+\tau\Delta\mathbf{t})$$

vanishes when $\tau = \tau_r$. Since F_r is differentiable,

$$\sum_{s=1}^{n} \frac{\partial F_r}{\partial x_s}\Delta x_s + \sum_{s=1}^{m} \frac{\partial F_r}{\partial t_s}\Delta t_s = 0,$$

each derivative being evaluated at $\mathbf{x}^r = \mathbf{x}+\tau_r\Delta\mathbf{x}$, $\mathbf{t}^r = \mathbf{t}+\tau_r\Delta\mathbf{t}$.

This is a system of n linear equations for $\Delta x_1, \Delta x_2, \ldots, \Delta x_n$; the determinant of the system has $F_{r,s}(\mathbf{x}^r; \mathbf{t}^r)$ as the element in the rth row and sth column, where $F_{r,s}(\mathbf{x}; \mathbf{t})$ denotes $\partial F_r/\partial x_s$. This determinant is a continuous function of the position of the n points $(\mathbf{x}^r; \mathbf{t}^r)$ on an open set in $X \times T$. Its value when the n points coincide at $(\mathbf{0}; \mathbf{0})$ is unity since it then reduces to the Jacobian. Therefore, if γ and δ are small enough, the determinant is bounded from zero. We can therefore solve the equations for the components Δx_r, in the form

$$\Delta x_r = \sum_{s=1}^{m} c_{r,s}\Delta t_s,$$

where the coefficients $c_{r,s}$ are continuous functions of $\mathbf{x}$, $\mathbf{t}$, $\Delta\mathbf{x}$ and $\Delta\mathbf{t}$.

In particular, if the only non-zero component of $\Delta\mathbf{t}$ is Δt_s,

$$\frac{\Delta x_r}{\Delta t_s} = c_{r,s}$$

and $c_{r,s}$ tends to a continuous limit as $\Delta t_s \to 0$. Therefore $\partial\alpha_r/\partial t_s$ exists and is continuous.

CHAPTER 9

FURTHER DEVELOPMENTS

83. Banach spaces

The applications of the theory of metric spaces in Chapter 8 were all in the fields of classical algebra and analysis. We turned various sets of 'points' into metric spaces by defining a 'distance'; but in each case, we assumed that the space had in addition a structure which was not part of the general theory of metric spaces. One line of development is to require that the initial abstract set always has a definite structure.

Let us start with an additive Abelian group, a set X of elements with the following properties:

(i) Every pair x and y of elements of X has a uniquely defined 'sum' denoted by $x+y$, and $x+y = y+x$.

(ii) Addition is associative; $(x+y)+z = x+(y+z)$.

(iii) There is a unique neutral element θ such that $x+\theta = x$ for every element x.

(iv) Every element x has a unique additive inverse, denoted by $-x$, such that $x+(-x) = \theta$.

An additive Abelian group X is called a *real vector space* if it admits the operation of multiplication by real numbers α, β, ..., an operation called scalar multiplication, with the following properties:

(i) For every real number α and for every element x of X, there is an element αx of X.

(ii) $(\alpha+\beta)x = \alpha x+\beta x, \alpha(\beta x) = (\alpha\beta)x$.

(iii) $\alpha(x+y) = \alpha x+\alpha y$.

(iv) If $\alpha = 1, \alpha x = x$.

These laws imply that $(-1)x = -x$ and that $(0)x = \theta$.

A *complex vector space* is similarly defined; the scalar multipliers $\alpha, \beta, \gamma, \ldots$ are then complex numbers. When we speak of a vector space, we may mean a real vector space or a complex vector space according to the context. A simple example of a real vector space is the set of all ordered triples of real numbers (x_1, x_2, x_3) with the rules

$$(x_1, x_2, x_3) + (y_1, y_2, y_3) = (x_1 + y_1, x_2 + y_2, x_3 + y_3)$$

$$\alpha(x_1, x_2, x_3) = (\alpha x_1, \alpha x_2, \alpha x_3);$$

this is ordinary vector algebra without scalar and vector products.

A normed vector space is a vector space on which is defined, for each point x of X, a real number called the norm of x and denoted by $\|x\|$, such that

(i) $\|x\| \geqslant 0$ and $\|x\| = 0$ if and only if $x = 0$.

(ii) For each scalar multiplier α and for each point x of X,

$$\|\alpha x\| = |\alpha| . \|x\|.$$

(iii) $\|x+y\| \leqslant \|x\| + \|y\|$.

For example, in the space just referred to, a possible norm is

$$\sqrt{(x_1^2 + x_2^2 + x_3^2)}.$$

A normed vector space can be metrized by taking

$$\rho(x, y) = \|x-y\|,$$

a metric which has the property $\rho(x, y) = \rho(x-y, \theta)$. In particular, $\rho(x, \theta) = \|x\|$ so that the norm of x is the 'distance of x from the origin'. If a normed vector space is complete with respect to the metric $\|x-y\|$, it is called a *Banach space*. The complex plane with norm $|z|$ is a Banach space.

The space of sequences $x = \{x_n\}$ of real or complex numbers is a vector space if

$$\{x_n\} + \{y_n\} = \{x_n + y_n\},$$

$$\alpha\{x_n\} = \{\alpha x_n\}$$

and θ is the sequence every member of which is zero. The following sequence-spaces are Banach spaces:

(i) The set m of all bounded sequences $x = \{x_n\}$ with norm

$$\|x\| = \sup |x_n|.$$

(ii) The set l^p of all sequences $x = \{x_n\}$ for which $\sum_1^\infty |x_n|^p$ is convergent, where $p \geqslant 1$, with norm

$$\|x\| = \left\{\sum_{n=1}^{\infty} |x_n|\right\}^{1/p}.$$

The space $C[a, b]$ of functions $x(t)$, continuous on $[a, b]$, becomes a vector space if we use the ordinary definitions of addition, zero and multiplication by a constant. If we use the norm

$$\|x\| = \sup\{|x(t)| : t \in [a, b]\},$$

$C[a, b]$ is a Banach space.

All these spaces were used in earlier chapters; but there was no need to introduce the term 'Banach space' as no use was made of the vector space property.

84. Hilbert spaces

In ordinary vector algebra on a real three-dimensional Euclidean space, there are two types of product, the scalar product and the vector product. If $\mathbf{x} = (x_1, x_2, x_3)$ and $\mathbf{y} = (y_1, y_2, y_3)$, the scalar product is

$$\mathbf{x} \cdot \mathbf{y} = x_1 y_1 + x_2 y_2 + x_3 y_3.$$

In particular, $\mathbf{x} . \mathbf{x}$ is the square of the length of the vector $\mathbf{x}$. In order to avoid confusion between multiplication of a vector by a scalar and the scalar product of two vectors, the scalar product of two vectors is often called their inner product.

Let X be any vector space. A complex valued function $\langle x, y\rangle$ of the ordered pair (x, y) of points of X is called an inner product if

(i) $\langle x, y\rangle$ is linear in x.
(ii) $\langle y, x\rangle$ is the complex conjugate of $\langle x, y\rangle$.
(iii) $\langle x, x\rangle \geqslant 0$.
(iv) $\langle x, x\rangle = 0$ if and only if $x = \theta$.

By (i), we mean that

$$\langle \alpha x_1 + \beta x_2, y\rangle = \alpha\langle x_1, y\rangle + \beta\langle x_2, y\rangle.$$

If we denote the conjugate of a complex number z by $\bar{z}$, (ii) then gives

$$\langle x, \alpha y_1 + \beta y_2\rangle = \overline{\langle \alpha y_1 + \beta y_2, x\rangle}$$
$$= \overline{\alpha\langle y_1, x\rangle + \beta\langle y_2, x\rangle}$$
$$= \bar{\alpha}\langle x, y_1\rangle + \bar{\beta}\langle x, y_2\rangle.$$

In particular

$$\langle \alpha x, y\rangle = \alpha\langle x, y\rangle, \quad \langle x, \alpha y\rangle = \bar{\alpha}\langle x, y\rangle;$$

hence, putting $\alpha = 0$,

$$\langle \theta, y\rangle = 0, \quad \langle x, \theta\rangle = 0.$$

Moreover it can be shown that

$$|\langle x, y\rangle|^2 \leqslant \langle x, x\rangle\langle y, y\rangle$$

with equality if and only if x and y are linearly dependent.

A vector space on which an inner product has been defined is called an *inner product space*. A trivial example is the set of all ordered triples $\mathbf{x} = (x_1, x_2, x_3)$ of complex numbers with

$$\langle \mathbf{x}, \mathbf{y}\rangle = x_1\bar{y}_1 + x_2\bar{y}_2 + x_3\bar{y}_3.$$

The set of all complex valued functions $x(t)$, continuous on a closed interval $[a, b]$ of the real line, is a complex vector space; if we write

$$\langle x, y\rangle = \int_a^b x(t)\,\overline{y(t)}\,dt,$$

it becomes an inner product space.

If, on an inner product space, we write

$$\langle x, x\rangle = \|x\|^2,$$

$\|x\|$ satisfies all the conditions for a norm; conditions (i) and (ii) are evident, and

$$\|x+y\|^2 = \langle x+y, x+y\rangle = \langle x, x\rangle + \langle x, y\rangle + \langle y, x\rangle + \langle y, y\rangle$$
$$= \|x\|^2 + 2\,\mathrm{rl}\,\langle x, y\rangle + \|y\|^2$$
$$\leqslant \|x\|^2 + 2|\langle x, y\rangle| + \|y\|^2$$
$$\leqslant \|x\|^2 + 2\|x\| . \|y\| + \|y\|^2$$

and so

$$\|x+y\| \leqslant \|x\| + \|y\|,$$

which is condition (iii).

An inner product space can be regarded as a metric space with metric $\rho(x, y)$ defined by

$$\rho(x, y) = \|x-y\|.$$

If the inner product space is complete with respect to this metric, it is called a *Hilbert space*.

The best known Hilbert space is the space of sequences $x = \{x_n\}$ of complex numbers such that $\sum_1^\infty |x_n|^2$ is convergent. It is a vector space if we define $\alpha\{x_n\} = \{\alpha x_n\}, \{x_n\}+\{y_n\} = \{x_n+y_n\}$. It is an inner product space if we define

$$\langle x, y\rangle = \sum_1^\infty x_n \bar{y}_n.$$

From this, we derive the norm

$$\|x\| = \sqrt{\sum_1^\infty |x_n|^2}.$$

And it is complete with respect to the metric $\|x-y\|$. This space was first studied by David Hilbert in his work on quadratic forms in infinitely many variables which he applied to the theory of integral equations.

Another example is the space of measurable functions $x(t)$ such that $|x(t)|^2$ is integrable in Lebesgue's sense over the closed interval $[a, b]$. We do not distinguish between two equivalent functions, two functions which differ only on a set of zero measure, so that a 'point' of the space is really an equivalence class of equivalent functions. With the obvious definitions of addition, multiplication by a scalar constant and zero, this is a vector space. The expression

$$\langle x, y\rangle = \int_a^b x(t)\,\overline{y(t)}\,dt$$

satisfies the conditions for an inner product. From this, we derive the norm

$$\|x\| = \sqrt{\int_a^b |x(t)|^2\,dt};$$

and the space is complete with respect to the metric $\|x-y\|$.

85. General topology

In §§ 83–84, we indicated two lines of development of the theory of metric spaces. In both, the abstract space from which we started was given greater specialization. Another line of development aims at freeing the subject from the idea of distance altogether.

In Chapter 3 we defined open and closed sets, which depended on the metric used. In some of the subsequent work, we could have dispensed with distance and used only closed and open sets. For example, we could have defined the continuity of a mapping $f: X \to Y$ by the condition that the inverse image of every open subset of Y is open.

Let X be an abstract space, and let $\mathcal{O}$ be a family of subsets of X with the following properties:

(i) The empty set $\varnothing$ and X itself are in $\mathcal{O}$.

(ii) The intersection of any two members of $\mathcal{O}$ is in $\mathcal{O}$.

(iii) The union of any number of members of $\mathcal{O}$ is in $\mathcal{O}$.

The family $\mathcal{O}$ is called a topology for X. The set X with a topology is called a topological space; and the members of the family $\mathcal{O}$ are called the open sets of the topological space. General topology is the study of the properties of topological spaces.

A metric space is a topological space, since the open sets of a metric space possess the above properties. A topological space is said to be metrizable if a metric can be defined on it such that the open sets determined by the metric are precisely the open sets which constitute the topology. General topology is a generalization of the theory of metric spaces since there are topological spaces which are not metrizable.

INDEX